T0142492

Advances in Intelligent Systems and Computing

Volume 838

The series "Advances in Intelligent Systems and Computing" contains publications on theory, applications, and design methods of Intelligent Systems and Intelligent Computing. Virtually all disciplines such as engineering, natural sciences, computer and information science, ICT, economics, business, e-commerce, environment, healthcare, life science are covered. The list of topics spans all the areas of modern intelligent systems and computing such as: computational intelligence, soft computing including neural networks, fuzzy systems, evolutionary computing and the fusion of these paradigms, social intelligence, ambient intelligence, computational neuroscience, artificial life, virtual worlds and society, cognitive science and systems, Perception and Vision, DNA and immune based systems, self-organizing and adaptive systems, e-Learning and teaching, human-centered and human-centric computing, recommender systems, intelligent control, robotics and mechatronics including human-machine teaming, knowledge-based paradigms, learning paradigms, machine ethics, intelligent data analysis, knowledge management, intelligent agents, intelligent decision making and support, intelligent network security, trust management, interactive entertainment, Web intelligence and multimedia.

The publications within "Advances in Intelligent Systems and Computing" are primarily proceedings of important conferences, symposia and congresses. They cover significant recent developments in the field, both of a foundational and applicable character. An important characteristic feature of the series is the short publication time and world-wide distribution. This permits a rapid and broad dissemination of research results.

More information about this series at http://www.springer.com/series/11156

Thouraya Bouabana-Tebibel (Deceased)
Lydia Bouzar-Benlabiod
Stuart H. Rubin
Editors

Theory and Application of Reuse, Integration, and Data Science

Editors
Thouraya Bouabana-Tebibel (Deceased)
Laboratoire de Communication dans les Systèmes Informatiques
Ecole nationale Supérieure d'Informatique
Algiers, Algeria

Lydia Bouzar-Benlabiod
Laboratoire de Communication dans les Systèmes Informatiques
Ecole nationale Supérieure d'Informatique
Algiers, Algeria

Stuart H. Rubin
Space and Naval Warfare Systems Center Pacific
San Diego, CA, USA

ISSN 2194-5357 ISSN 2194-5365 (electronic)
Advances in Intelligent Systems and Computing
ISBN 978-3-319-98055-3 ISBN 978-3-319-98056-0 (eBook)
https://doi.org/10.1007/978-3-319-98056-0

Library of Congress Control Number: 2018951100

This Springer imprint is published by the registered company Springer Nature Switzerland AG
The registered company address is: Gewerbestrasse 11, 6330 Cham, Switzerland

This book is dedicated to the memory of
Professor Thouraya Bouaban-Tebibel.

Preface

The goal for this book is to survey recent trends in reuse and integration by way of presenting seven selected works, which together cover the recent developments. Drs. Bouzar-Benlabiod and Rubin would first like to avail themselves of the opportunity to thank the late Professor Bouabana-Tebibel—without whom there would be no book. Her foresight into the need to compile a book such as this one was not only insightful, but inspirational.

Reuse is not merely limited to code reuse. Reuse is properly incorporated into every software development step for various domains—from robotics and security authentication to issues of environment and so on. The present challenge pertains to the adaptation of solutions from one language to another or from one domain to another. It is critical that reused artifacts be validated in each environment in which they are deployed.

Robots are becoming ubiquitous, «Improved Logical Passing Strategy and Gameplay Algorithm for Humanoid Soccer Robots using Colored Petri nets» uses robots that must be successfully programmed with image recognition, positioning, ball kicking, and a playing strategy in order to successfully obtain a match. The improved Passing with Logical Strategy (iPaLS) is a defined Colored Petri Net (CPN) modeling, and simulation is used in order to test the various scenarios of a system that implements iPaLS and helps to prove the advantages of this algorithm over a strategy of having each NAO robot kicks the ball toward the goal without regard for their teammates. Robots are also largely used at home. Simulations and filming their behavior are still the most frequently used techniques to analyze them. «Analyzing Cleaning Robots using Probabilistic Model Checking» reuses Probabilistic Model Checking (PMC) to perform such an analysis. Using PMC, we can verify whether a robot trajectory (described in terms of an algorithm), satisfies specific behaviors or properties (stated as temporal formulas). For instance, one can measure energy consumption, time to complete missions, etc. As a consequence, we can also determine whether an algorithm is superior to another in terms of such properties. We choose the PRISM language; it can be used in more than one PMC tool. We also propose a DSL to hide the PRISM language from the end user; this DSL passes an automatic sanity test. Programs and properties written in the proposed DSL are automatically translated into the notation of PRISM by following a set of mapping rules. An integrated environment is proposed to support the authoring of the algorithms, properties in the proposed DSL, and checking them in a Probabilistic Model Checker. Finally, we present an evaluation of three algorithms defining our proposal.

Another aspect of reuse is the modeling languages, such as UML. UML is a semi-formal notation largely adopted in the industry as the standard language for software design and analysis. Its imprecise semantics prevents any verification task. However, a formal semantics can be given to UML diagrams, for instance, through their transformation to models having a formal semantics, such as CPNs. CPNs are an efficient languages for UML state machine formalization and analysis. In order to assist the UML modeler in understanding the report generated by the Petri net tool, a method to construct UML diagrams from the returned report is given in «From Petri Nets to UML: A new approach for model analysis».

Environmental issues also need reuse paradigms. Automated negotiation is used as a tool for modeling human interactions with the aim of making decisions when participants have conflicting preferences. Although automated negotiation is extensively applied in different fields (e.g., e-commerce), its application in environmental studies remains unexplored. «Using Belief Propagation-based Proposal Preparation for Automated Negotiation over Environmental Issues» focuses on negotiation in environmental resource management projects. The primary objective of the study is to reach agreement/conclusion as fast as possible. In order to achieve this objective, an agent-based model with two novel characteristics is proposed. The first is for automating the process of proposal offering using Markov Random Fields and belief propagation (BP). The second is the ability to estimate stakeholders preferences through an argument handling (AH) model. These experiments demonstrated that the AH model and the BP-based proposal preparation (BPPP) approach improve the performance of the negotiation process. A combination of these two modules outperforms the conventional utility-based approach by decreasing the number of negotiation rounds up to 50 percent. Another work in the field of environment is presented herein. It concerns real-time fault diagnosis for streaming vibration data from turbine gearboxes. Failure of a wind turbine is largely attributed to faults that occur in its gearbox. Maintenance of this machinery is very expensive, mainly due to large downtime and repair costs. While much attention has been given to detect faults in these mechanical devices, real-time fault diagnosis for streaming vibration data from turbine gearboxes is still an outstanding challenge. Moreover, monitoring gearboxes in a wind farm with thousands of wind turbines requires massive computational power. «SAIL: A Scalable Wind Turbine Fault Diagnosis Platform A Case Study on Gearbox Fault Diagnosis» proposes a three-layer monitoring system: sensor, fog, and cloud layers. Each layer provides a special functionality and runs part of the proposed data-processing pipeline. In the sensor layer, vibration data is collected using accelerometers. Industrial single-chip computers are the best candidates for node computation. Since the majority of wind turbines are installed in harsh environments, sensor node computers should be embedded within wind turbines. Therefore, a robust computation platform is necessary for sensor nodes. In this layer, we propose a novel feature extraction method, which is applied over a short window of vibration data. Using a time-series model assumption, our method estimates vibration power at high resolution and low cost. Fog layer provides Internet connectivity. Fog server collects data from sensor nodes and sends it to the cloud. Since many wind farms are located in remote locations, providing network connectivity is challenging and expensive. Sometimes a wind farm is offshore, and a satellite connection is the only solution. In this regard, we use a compressive sensing algorithm by deploying them on fog servers to conserve communication bandwidth. The cloud layer performs most computations. In the online mode, after decompression, fault diagnosis is performed using a trained classifier, while generating reports and logs. When, in the off-line mode, model training for a classifier, parameter learning for feature extraction in the sensor layer, and dictionary learning for compression on fog servers and decompression are performed. The proposed architecture monitors the health of turbines in a scalable framework by leveraging distributed computation techniques. The proposed empirical evaluation of vibration datasets, obtained from real wind turbines, demonstrates high scalability and performance in diagnosing gearbox failures, i.e., with an accuracy greater than 99%, for application in large wind farms.

Cloud computing enables the outsourcing of big data analytics, where a third-party server is responsible for data management and processing. A major security concern of the outsourcing paradigm is whether the untrusted server returns correct results. «Efficient Authentication of Approximate Record Matching for Outsourced Database» considers approximate record matching in the outsourcing model. Given a target record, the service provider should return all records from the outsourced dataset that are similar to the target. Identifying approximately duplicate records in databases plays an important role in information integration and entity resolution. In this paper, we design ALARM, an authentication solution of outsourced approximate record matching to verify the correctness of the result. The key idea of ALARM is that besides returning the similar records, the server constructs the verification object (VO) to prove their authenticity, soundness, and completeness. ALARM consists of four authentication approaches, namely V S 2, E-V S 2, G-V S 2, and P -V S 2. These approaches endeavor to reduce the verification cost from different aspects. We theoretically prove the robustness and security of these approaches and analyze the time and space complexity for each approach. We perform an extensive set of experiments on real-world datasets to demonstrate that ALARM can verify the record-matching results at minimal cost.

«Active Dependency Mapping A Data-Driven Approach to Mapping Dependencies in Distributed Systems» introduces Active Dependency Mapping (ADM), a method for establishing dependency relations among a set of interdependent services. Consider the old expression among sysadmins: The way to discover who is using a server is to turn it off and see who complains. ADM takes the same approach to dependency mapping. We directly establish dependency relationships between entities in an ecosystem of interacting services by instrumenting a user facing application, and systematically blocking access to other entities, observing which blocks impact performance.

Our approach to ADM differs in an important way from "turning it off to see who complains." Although ADM depends on degrading the network environment, our goal is to minimize the impact of traffic blocks by reducing the duty cycle to determine the shortest block that enables dependency mappings. In an ideal case, dependencies may be established with no one complaining. Artificial degradation of the network environment could be transparent to users; run continuously, it could identify dependencies that are rare, or occur only at certain timescales. A useful by-product of this dependency analysis is a quantitative measurement of the resilience and robustness of the system. This technique is intriguing for hardening both enterprise networks and cyber physical systems.

As a proof of concept experiment to demonstrate ADM, we have chosen the Atlassian Bamboo continuous integration and delivery tool used at Lincoln Laboratory—together with associated services, as an ecosystem whose dependencies need to be mapped. Using the time necessary to complete a standard build as one observable metric, data on the actual dependencies of the Bamboo build server is obtained by selectively blocking access to other services in its ecosystem and is presented. Current limitations and suggestions for further work are discussed.

This book includes high-quality research papers, written by experts in information reuse and integration, and covers some of the most recent advances in the field. These papers are extended versions of the best papers, which were presented at the IEEE International Conference on Information Reuse and Integration (IRI) and the IEEE International Workshop on Formal Methods Integration (FMI), which was held in San Diego, California, in August 2017.

Lydia Bouzar-Benlabiod
Stuart H. Rubin

Contents

Improved Logical Passing Strategy and Gameplay Algorithm for Humanoid Soccer Robots Using Colored Petri Nets

Kieutran Theresa Pham, Chelsea Cantone, and Seung-yun Kim(✉)

Department of Electrical and Computer Engineering,
The College of New Jersey, 2000 Pennington Road, Ewing, NJ 08618, USA
{phamk2, cantoncl, kims}@tcnj.edu

Abstract. RoboCup, a project originally named the Robot World Cup Initiative, challenges people around the world to program robots that are capable of competing in a soccer tournament. The goal is that one day, a group of robots will be able to match the playing ability of a human soccer team, and even be able to win against humans in soccer. The RoboCup Standard Platform League utilizes teams of humanoid NAO robots. These robots must be successfully programmed with image recognition, positioning, ball kicking, and a playing strategy in order to successfully get through a match. In human soccer, a team strategy is crucial to winning a match, but not all RoboCup teams have programmed their team strategies to call on the robots to work together in order to score a goal. The improved Passing with Logical Strategy (iPaLS) is an algorithm that proposes passing of the ball between players to more quickly score a goal. This algorithm is an extension of the Passing with Logical Strategy (PaLS) algorithm, which proposed a more rudimentary method of passing between players. iPaLS builds upon PaLS by further exploring the kicking decisions that must be made by the NAO robot and considers ways to hinder the opposing team's ability to gain possession of the ball and ways to regain possession of the ball if possession is lost. Colored Petri net modeling and simulation is used in order to test the various scenarios of a system that implements iPaLS and helps to prove the advantages of this algorithm over a strategy of having each NAO robot kick the ball towards the goal without regard for their teammates.

Keywords: Petri nets · Colored Petri nets · Robot soccer · RoboCup NAO

1 Introduction

The field of robotics has advanced rapidly in the past decade. Each year robots become more intelligent and more capable of completing complex tasks. Engineers and scientists in all domains look for new ways to improve and develop a humanoid robot's abilities to emulate human behavior, finding new scenarios of application for robotics. For example, concept learning has been applied in a multi-agent system to improve the environment perception and communication [1]. The integration of robotics into human

T. Bouabana-Tebibel et al. (Eds.): IEEE IRI 2017, AISC 838, pp. 1–22, 2019.
https://doi.org/10.1007/978-3-319-98056-0_1

life depends on the autonomy of the robot. A humanoid robot is based on the human body used in studying bipedal locomotion.

Humanoid NAO robotic platforms [2], for example, have been used in many different applications, from simple educational tools [3] to object recognition using a modified simultaneous recurrent network [4]. In order to promote robotics and artificial intelligence (AI) research, an annual Robot Soccer World Cup was founded [5]. The RoboCup Initiative works to foster artificial intelligence in robotics by providing the domain of soccer to challenge people to integrate and explore new technologies and algorithms. The RoboCup has a range of leagues, including the Standard Platform League (SPL), in which the NAO robot is the standard technology for the games. Years later, much progress has been made towards that goal, but there still exists many milestones in research and development to be done before robots will be near able to play against human soccer players.

During a soccer match, the human brain is able to process images, make decisions, call on its body to act instantly, and keep a strategy in mind all at the same time. Robots are a long way from having these capabilities to the same speed as that of a human. The RoboCup SPL League currently requires the use of humanoid NAO robots for its matches, and researchers around the globe work to program teams of these robots to engage in a soccer match to the best of their abilities using collaboration and decision making methods. A study [6] claimed that a collaboration process needs enormous information sharing which requires a reliable decision process framework. Many teams have been able to get the NAOs to process inputs such as images, allowing them to detect objects and their surroundings, as well as understand how to kick the soccer ball when in proximity. What some teams must improve upon is a framework of programming the robots to think like a human – to play with a strategy involving teamwork.

Collaboration within multi-agent systems is highly dependent on how well each agent can effectively share and upgrade their conceptualizations of their environment. A main focus of analyzing multi-agent systems is development of solutions to the concept learning problem, where individual agents are trying to learn a concept and teach it to another agent [1]. In recent years RoboCup SPL teams have made improvements to team communication to improve learning between agents, for example, a deep learning approach using convolutional neural networks [7]. The 2016 Champion RoboCup SPL Team, UPennalizers, implemented a system to crowdsource the ball's location on the field and vote where it is most likely located [8], similar to how Team rUNSWift utilizes a method to team-track the ball across the field [9]. Similarly, in the RoboCup Simulation League focusing on algorithm and strategy design, the 2014 team UT Austin Villa found success in implementing "kick anticipation" and broadcasting their targets to their team, much like how a human soccer team would communicate where they are kicking the ball [10]. This allows teammates to then anticipate where the ball may end up so it may be received further down the field. It is a continuing initiative to model and simulate team coordinating strategies.

To help determine how the NAOs should make decisions and act upon them, a Petri net (PN) can be used to analyze a newly developed algorithm. A majority of systems and processes can be characterized as concurrent and distributed in their inherent operation, relying on communication, synchronization, and resource sharing to be successful. A PN is a graphical and mathematical modeling tool that can be used to describe and analyze these complex systems and processes. Many researchers have been using PNs to help visualize the problems, ranging from directing complicated traffic at a train station [11] to practicing collaborative application procedures based on user preferences [12]. Petri nets were also used to model a methodology of controlling autonomous robots [13], as well implementing a generic hybrid monitoring approach for autonomous mobile robots [14].

Many extended Petri nets have also been proposed to accommodate the varying needs of different researchers, as regular Petri nets can be limited in their capabilities. Fuzzy Petri net modeling and simulation were exploited to develop algorithms for a self-navigating firefighting robot [15]. A Distributed Agent-Oriented Petri net was used to model the methodology of controlling autonomous robots [13], and a Modified Particle Petri net was developed to implement the generic hybrid monitoring approach [14].

With robots flourishing at every turn, the need for sophisticated modeling has become increasingly pertinent. Colored Petri nets (CPNs) were introduced [16], incepting a programming language-based Petri net. CPNs incorporate concepts of color sets for different types of data, combine modeling with programming language, and introduce ideas of hierarchy. Because of these added capabilities, CPNs allow for the design of a more specific and logical model than what a Petri net alone is capable of [17]. There are many examples of the use of CPNs in robotics fields. CPNs were used to model the behavior of a human-wearable robot, where robotic limbs were developed that attached to a human to coordinate activity with its wearer to reduce human workload [18]. CPNs were also used to model a mechanism for an autonomous robot boat team to smoothly handle human interrupts [19]. In this work, an operator must intervene to prevent unforeseeable harm to the robot team. The colored tokens were used to represent the complex and hierarchical team-planning structure, signifying the specific roles of teammates, and encoding two types of interrupts.

By modeling a system or algorithm using a CPN, the system or algorithm can be more closely analyzed for methods of improvement and increased efficiency. This journal presents an algorithm that was improved with the help of Color Petri net modeling. That is why the CPN was chosen to model and simulate a team-centric algorithm that proves the increased effectiveness a teamwork based algorithm would bring to a RoboCup team. These models can also be easily built upon or modified, as changes and improvements are made to a system. In this case, a team-playing algorithm from a previous work was enhanced and built upon to develop a stronger algorithm.

The remainder of the paper is organized as follows: Sect. 2 presents background information on the fundamental concepts and theory for Petri nets and Colored Petri nets that were essential to the development of the modeled algorithm. Section 3 explains the realization of the proposed algorithm and its development, beginning with an explanation of the previously proposed PaLS algorithm, and then describing its transition into iPaLS. Section 4 starts off by showcasing the possible application of the PaLS algorithm through simulation of a CPN by examining the efficiency of the algorithm when compared with a path optimization algorithm based on fuzzy and genetic logic. The section then unveils the model for the iPaLS algorithm, followed by its own simulation results. Section 5 gives a summary of the completed work and its implications, and then concludes the piece with a discussion on future work to be done.

2 Background

2.1 Petri Nets

A Petri net (PN) is a directed bipartite graph with two different types of nodes – places, drawn as ellipses, and transitions, drawn as rectangles. It consists of a distribution of tokens, drawn as dots, which are positioned on the places. For modeling, places represent conditions, transitions represent events, and tokens represent the truth of a condition. The tokens are used in these nets to simulate the dynamic and concurrent activities of systems. With the execution of a PN, occurrences of enabled transitions remove tokens from input places and add tokens to output places as allowed by integer arc weights and based on the transition enabling and firing rule [17]. The rule follows that a transition is enabled if each input place of a transition holds tokens that equal the weights of the connecting arcs. If it fires, the net removes the number of tokens equal to the weights of the arcs from their connecting input places to the transition, and brings tokens to output places that equal the weight of the arcs from a connecting transition. The different states that a Petri net goes through are known as markings. Table 1 shows the formal definition of a PN [20] while an example of a PN in its initial marking is shown in Fig. 1.

Table 1. Formal definition of a Petri net

A Petri net is a 5-tuple, $PN = (P, T, F, W, M_0)$ where:
$P = \{p_1, p_2, \ldots, p_m\}$ is a finite set of places, $T = \{t_1, t_2, \ldots, t_n\}$ is a finite set of transitions, $F \subseteq (P \times T) \cup (T \times P)$ is a set of arcs (flow relation), $W: F \rightarrow \{1,2,3,\ldots\}$ is a weight function, $M_0: P \rightarrow \{0,1,2,3,\ldots\}$ is the initial marking. $P \cap T = \phi$ and $P \cup T \neq \phi$.
A Petri net structure $N = (P, T, F, W)$ without any specific initial marking is denoted by N.
A Petri net with the given initial marking is denoted by (N, M_0).

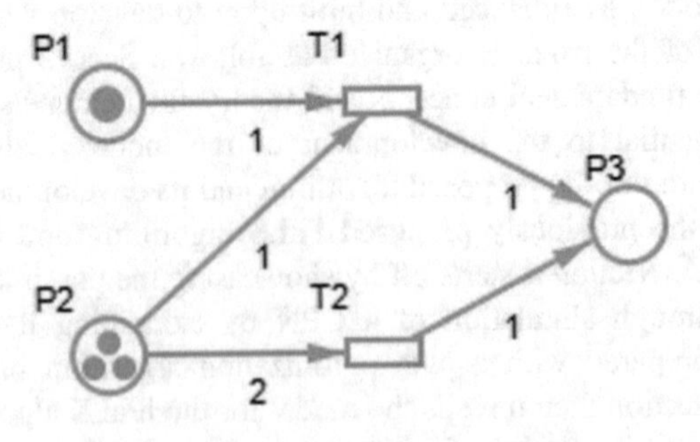

Fig. 1. Petri net example

The example shows that places can hold different amounts of tokens, and arcs have weights, which control the amount of tokens that can pass from a transition to a place. If the given PN example is simulated, the result is the emptying of tokens from P1 and P2, and placing two tokens into P3.

Petri nets using only places and transitions do not scale well to large systems, and it is hard to manipulate and differentiate data or create a hierarchy. The development of a programming language-based Petri net made more functionality possible in order to allow for better modeling of concurrent and distributed systems.

2.2 Colored Petri Nets

The Colored Petri net (CPN) model allows the user to create tokens of different data types, referred to as color sets, and combines the Petri net with known functionalities from programming languages based on standard meta language (SML). Unlike PNs, CPNs can store and associate data with their tokens, and arc weights must include the matching data type of the tokens in order for transitions to be enabled and fired. The transition enabling and firing rule is the same as with PNs, though on top of arc integer weights, arcs can also hold expressions and functions, and transitions can include guards. A formal definition of a non-hierarchal Colored Petri net is provided in Table 2 [16].

Table 2. Formal definition of a Colored Petri net

A Colored Petri net is a 9-tuple, $CPN = (P, T, A, \Sigma, V, C, G, E, I)$ where:

1. P is a finite set of places.
2. T is a finite set of transitions such that $P \cap T = \emptyset$.
3. $A \subseteq P \times T \cup T \times P$ is a set of directed arcs.
4. Σ is a finite set of non-empty color sets.
5. V is a finite set of typed variables such that $Type[v] \in \Sigma$ for all variables $v \in V$.
6. $C: P \to \Sigma$ is a color set function that assigns a color set to each place.
7. $G: T \to EXPR_V$ is a guard function that assigns a guard to each transition t such that $Type[G(t)] = Bool$.
8. $E: A \to EXPR_V$ is an arc expression function that assigns an arc expression to each arc a such that $Type[E(a)] = C(p)_{MS}$, where p is the place connected to the arc a.
9. $I = P \to EXPR_\emptyset$ is an initialization function that assigns an initialization expression to each place p such that $Type[I(p)] = C(p)_{MS}$.

CPNs extend the original Petri net in multitudes of ways, allowing for more complex logic in modeling and more specific representations in token values. Figure 2 shows a simple Colored Petri net example created in CPN Tools [21], the modeling software that is used for this paper. CPNT is a software platform for editing, simulating, and analyzing CPNs. Each place is labeled with a data type, known as a color set, in all caps outside of the place, along with a label of how many tokens are in the place and what the tokens hold in the format "n`data" where n is the number of tokens followed

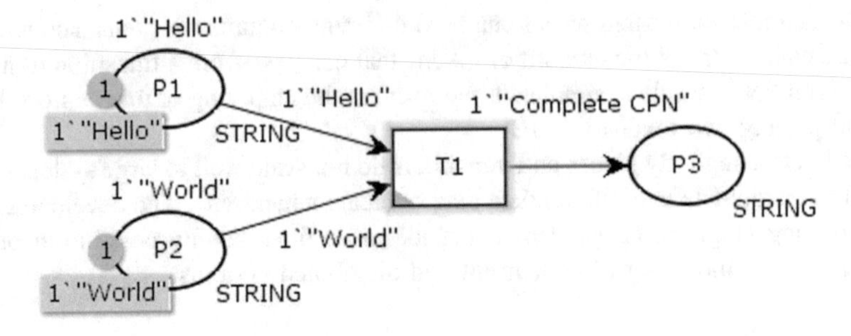

Fig. 2. Colored Petri net example

by the data that the tokens hold, which must match the color set defined for that place. The example shows two places which initially hold tokens of the STRING color set that will pass through the arcs when the transition is fired, since the arc weights match the tokens. This then outputs the string "Complete CPN" to place P3. Place P1 holds one token with a string that says "Hello" while place P2 holds one token with a string that says "World". Arcs hold weights which limit what can be sent from a place to a transition, and in CPN Tools, weights include not only the number of tokens but also the data held by the tokens. When the transition fires, it will send a token with a string that says "Complete CPN", which will end up in place P3.

3 Inception and Development of Proposed Algorithm

3.1 Previous Work

Various methodologies have been proposed and implemented for NAOs to play soccer in RoboCup effectively. One team from the University of Texas at Austin examined the problem of the complicated vision system required for NAO robots to compete in RoboCup [22]. An auto-calibration system was utilized to take the image from the NAO and tune it to match that of a static image taken from a high quality camera for color matching. The vision system utilizes blob formation, object detection, and transformation to recognize the objects and shapes surrounding the NAO robot [23].

In terms of soccer with human players, the issue of real soccer ball detection against white lines and other players was investigated [24]. Authors created a vision system that used elliptical Gaussian kernel to detect lines to eliminate them from the image, and used blob detection to eliminate blobs larger than the ball. These are just a small number of details that must be considered while working towards well-functioning robotic soccer players with a comprehensive vision system. The object detection was performed in a dynamic environment for the RoboCup competitions with 3D depth information using a Kinect and colored blob detection provided by the UAVision library [25]. Using this proposed method, the goalie robot was able to trace and eventually predict the real-time trajectory of a ball.

To identify pertinent objects from the foreground, different approaches have been proposed to perform real-time object segmentation. A distance-based algorithm using an RGB-D camera was proposed to encode depth information into the recorded images utilizing PointCloud filtering and RANSAC based plane fitting to remove unnecessary data outliers [26]. Instead of detecting object edges, the feature extraction used two different color spaces, i.e., HSV and YUV, and an omnidirectional camera to achieve object segmentation [27]. In this method, color thresholds were set for each of the different items in play, such as the robot, ball, or soccer net in order to perform the object segmentation. Another approach explores the real-time performance benchmarks of the OpenCV framework for computer vision on mobile platforms to evade the processing power limitation [28].

3.2 PaLS Algorithm

Unfortunately, what several of these algorithms fail to account for in their strategies is passing between players, a major aspect of gameplay in real soccer. The previously proposed algorithm, Passing with Logical Strategy (PaLS) focuses on optimizing gameplay by passing to nearby ally NAOs clear of opponent NAOs in the ball's pathways [29]. This method of navigating through a game using a teammate based algorithm was investigated and modeled using Colored Petri nets.

For ease of visualizing and modeling the algorithm, the soccer field was partitioned into a 12×3 coordinate system shown in Fig. 3. This system is useful as the penalty box is enclosed in a single coordinate space (12, 2), providing a place on the coordinate system for the NAOs to aim for when intending to score a goal, as well as providing x-values that easily mark the position of the allies and opponents across the width of the field throughout the game. For the start of the game, the ally NAOs are placed at x = 6 while the opponent NAOs start at x = 7.

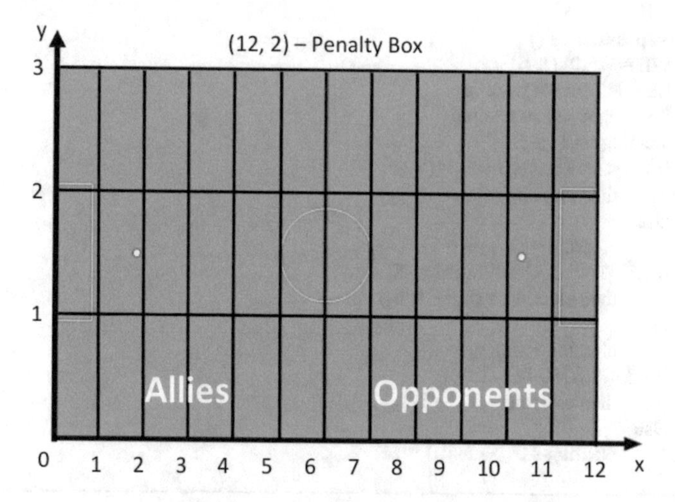

Fig. 3. Coordinate representation of the soccer field

The logic of the PaLS algorithm is shown in Fig. 4, which directs the NAO robots to analyze their position, based on the 12 × 3 coordinate system. When in possession of the ball, based on the NAO's current position, it will decide to either pass to a teammate or make a goal shot.

PaLS Algorithm

```
1:  if x_b < 10
2:        decideDirection()
3:  else
4:        if (x_b, y_b) = (12,1) OR (x_b, y_b) = (12,3),
5:              then (x_b, y_b) = (11, y_b )
6:        else
7:              if clear = true,
8:                    then attempt goal shot
9:              else
10:                   decideDirection()
11:
12: function decideDirection() {
13:       a_i = ally
14:       o_j = opponent
15:       i, j = {1, 2}
16:       1 = first detected
17:       2 = second detected
18:       doa1 = distance between a_1 and o_1
19:       doa2 = distance between a_2 and o_2
20:       if doa1 > doa2
21:             then doa1Shorter = true AND positioning()      AND      (x_b, y_b) =
(x_a1, y_a1)
22:       else
23:             positioning() AND (x_b, y_b) = (x_a2, y_a2)
24: }
25:
26: function positioning () {
27:       WL = {(*,0),(*,4)}
28:       LS = left sensor reading
29:       RS = right sensor reading
30:       clearing = (x_c, y_c)
31:       if RS < 1< LS AND WL = (x_a,0)
32:             then clearing = (x_a − 1, y_a)
33:       else
34:             clearing = (x_a, y_a + 1)
35:       if LS <1 <RS AND WL = (x_a,4)
36:             then clearing = (x_a − 1, y_a)
37:       else
38:             clearing = (x_a, y_a − 1)
39:       if LS < 1 AND RS< 1
40:             then clearing =(x_a − 1, y_a)
41:       else
42:             clearing =(x_a + 1, y_a)
43: }
```

Fig. 4. Passing with logical strategy (PaLS) algorithm

The PaLS algorithm has NAOs initially check to see if the x position of the ball, x_b, is close enough to the goal to attempt scoring. If $x_b < 10$, PaLS directs the NAO in possession of the ball to pass to an ally NAO in the best position to receive the ball, (x_a, y_a), using the decideDirection function. This function analyzes the distances between two allies and their closest opponents, *doa1* and *doa2*, and determines if the Boolean *doa1Shorter* is true to help it decide who to pass to.

Otherwise, if x_b is close enough to the goal but the NAO is not in the 12th x position, and there are no opponents in ball's path aside from the opposing goalie (determining the *clear* Boolean variable), the robot can attempt to score. If the ball is in the 12th x position (line 4), the NAO will kick the ball into the 11th x position (line 5) for a better shot at the goal. If there are opponents in the way, the NAO will pass to another NAO by calling on the decideDirection function. This will be used for a majority of the game.

The positioning function called in decideDirection locates the optimized spaces (clearings) around the NAO to make a successful pass to an ally. Throughout gameplay in RoboCup, the NAO is capable of perceiving its environment through sensory information taken from its two cameras, tactile sensors, pressure sensors, four microphones, and two chest ultrasonic sensors. The two ultrasonic transmitters and two receivers, have a resolution of 1 cm–4 cm and a detection range of 0.20 m–0.80 m. Each sensor covers an effective cone of 60°. The ultrasonic sensor readings of the NAO are labeled as *LS* (left senor) and *RS* (RS) with 0.20 to 0.80 m of reasonable range to detect allies and opponents. In conjunction, white lines (WL) are used to approximate allowable locations on the field, as robots should not move out of bounds. The field partitioned into 2 m by 0.75 m boxes will contain the location of each ally in respect to one another. Positioning will optimize the robot's location on the field by locating clearings around the robot free of opponents and white lines. Once the robot has defined the clearings around them, they are able to reposition themselves to increase their range of unobstructed movement (i.e. move away from opponents before passing).

3.3 Improved PaLS (iPaLS) Algorithm

The PaLS algorithm [28] centered around the use of a 12 × 3 coordinate system, where the location in the coordinate system was used as a means of determining whether the NAO robot was close enough to the goal or not to score. In the long term though, the use of a coordinate system will not be the most effective way of determining proximity to an object. One box in the 12 × 3 system is 0.75 m × 2 m, which is a large area to use for characterizing the location of a NAO robot, which only takes up approximately 0.3 m × 0.2 m space standing, and if the size of the RoboCup field changes and grows, the method becomes less and less effective to the point where a new coordinate system would eventually need to be made.

The distance between the goal box and any player is variable as players move about the field. Though, there exists a minimum distance that the NAO needs to be from the goal to be able to score. This distance can be calculated through the use of image recognition in OpenCV. Goal score attempts can then be made based upon calculated distance from the goal and the NAO determining whether they are close enough. If they are, they can set a boolean variable, *close*, to true. A local binary pattern (LBP) classifier can be used to train the robots ahead of time on recognizing the ball, other NAO robots, and the goal [7]. The robots can then be calibrated to recognize the distance between itself and other objects using a triangle similarity, knowing the actual width of the ball and comparing the perceived image with an image and distance that the robot already recognizes. Pseudocode for determining the distance using OpenCV is shown below (Fig. 5).

Determining Distance

```
1:    int knownWidth, knownDistance
2:    int focalLength, perceivedWidth
3:    int distance
4:
5:    camera = cv2.VideoCapture(NAOCamera)
6:
7:    while True
8:          image = camera.read()
9:          perceivedWidth = findObject(object);
10:         focalLength =                    (perceivedWidth*knownDistance)/knownWidth
11:
12:         distance = (actualWidth*focalLength)/
      perceivedWidth
13:         if(distance < kickingThreshold
14:               close = true;
15:
16:   findObject(object) {
17:         use trained classifier to recognize a given object     from a set of objects that the robot
has been trained to      recognize
18:   }
```

Fig. 5. Pseudocode for determining distance using OpenCV

If *close* is true, the NAO robot will then check if the path is clear before attempting to score a goal. The coordinate system is currently still used as a means of keeping track of the general location of the NAO for purposes of the CPN simulation, though beginning to incorporate concepts of image processing will create a clearer view of the system and its true implementation for NAOs when passing the ball and scoring.

Aside from exploring the soccer-playing algorithm with image processing, additional changes have been made to strengthen the team play strategy of the algorithm. When PaLS was developed, opponents were not factored in. In the scenario where the ally NAOs lost possession of the ball, there was no strategy for getting it back. iPaLS, the improved Passing with Logical Strategy algorithm, now assigns action to the robots in the case that ball possession is lost. Rather than having all of the offensive players move in the direction of the opponents to try and get the ball, only two of the offensive players, A2 & A3, will move to attempt to get the ball back while one offensive player, A1, will remain the position it was at when possession was lost. There is also the addition of a defense player for the allies to help them get the ball back, a player who would be stationary on the ally side of the field until opponent players gain possession of the ball. In those cases, the defense ally would then take action, moving to try and

iPaLS Algorithm

```
1:  while gameplay
2:      if allyPossesion
3:          close = searchGoal()
4:          if !close
5:              decideDirection()
6:          else
7:              if clear
8:                  then attempt goal shot
9:              else
10:                 decideDirection()
11:     else
12:         A2 & A3 = (x_a − 1, y_a)
13:         A1 = (x_a, y_a)
14:         searchBall()
15:
16: function decideDirection() {
17:     a_i = ally
18:     o_j = opponent
19:     i, j = {1, 2}
20:     1 = first detected
21:     2 = second detected
22:     doa1 = distance between a_1 and o_1
23:     doa2 = distance between a_2 and o_2
24:     if doa1 > doa2
25:         then doa1Shorter = true AND positioning() 26: AND (x_b, y_b) =
(x_{a_1}, y_{a_1})
27:     else
28:         positioning() AND (x_b, y_b) = (x_{a_2}, y_{a_2})
29: }
30:
31: function searchGoal(){
32:     use LBP classifier to find goal in image
33: }
34: function searchBall() {
35:     use LBP classifier to find goal in image
36: }
37:
38: function positioning () {
39:     WL = {(*,0),(*,4)}
40:     LS = left sensor reading
41:     RS = right sensor reading
42:     clearing = (x_c, y_c)
43:     if RS < 1< LS AND WL = (x_a,0)
44:         then clearing = (x_a − 1, y_a)
45:     else
46:         clearing = (x_a, y_a + 1)
47:     if LS <1 <RS AND WL = (x_a,4)
48:         then clearing = (x_a − 1, y_a)
49:     else
50:         clearing = (x_a, y_a − 1)
51:     if LS < 1 AND RS< 1
52:         then clearing =(x_a − 1, y_a)
53:     else
54:         clearing = (x_a + 1, y_a)
55: }
```

Fig. 6. Improved passing with logical strategy (iPaLS) algorithm

regain possession for the allies, passing the ball back to one of the offensive allies if the chance presents itself.

PaLS did not assign action in the scenario that ball possession was lost and also did not involve the assignment of offensive versus defensive players. Methods for implementation of the vision system had also not been seriously considered yet during the development of PaLS. These additions made in the iPaLS algorithm are tantamount to ensuring the NAO robots' ability to handling different scenarios that could be encountered as well as strengthening the gameplay strategy. The assignment of players to such positions as offensive and defensive also allows the gameplay of the robots to more closely emulate that of a human soccer game. These additions will be more evident in the CPN model shown later. Figure 6 outlines the algorithm in pseudocode that a single ally NAO will utilize throughout gameplay.

iPaLS gives a fuller algorithm for gameplay that considers many scenarios that PaLS had yet to breach. It also begins to give a more realistic implementation for the robot with the intended usage of image processing when searching for the goal and the ball. In future work, more image processing will be explored to make all parts of the strategy that are currently dependent on the coordinate system, such as the passing mechanism, dependent on image processing instead.

4 Modeling and Testing of Algorithms

CPN Tools allowed for the fruition of the PaLS algorithm as a Colored Petri net that can be simulated to observe the results of the algorithm in application. It has continued being used as the means of modeling the iPaLS algorithm. By modeling an algorithm as a CPN, the gameplay optimization can be proven through the results of the simulations.

4.1 Modeling of PaLS

In creating the CPN for PaLS, rather than creating a model to match the exact coordinate system, a model was created that outlined the decisions and actions that the robots would run through during a game when it came into possession of the ball. The coordinate system was used to help follow the positions of the NAOs and determine when to attempt a goal shot in the model as represented in the algorithm. The model, shown in Fig. 7, showcases the scenarios where the ball can go from midfield on after the ally team does kickoff.

A self-created color set LOC, a set of two integers that represent the ball's coordinates, was assigned to all places and carried through the model to represent the coordinate location of the players as the game progressed. The initial marking of the model shows the kickoff positions of the three offensive ally players. The NAO in control of the ball, based on the distances between the other two allies and the closest opponents, will decide which ally to pass to. This is represented by the if statement in the arc expressions from the transition "Decide Direction", which is intended to mimic the decideDirection function from the PaLS algorithm shown in Fig. 4.

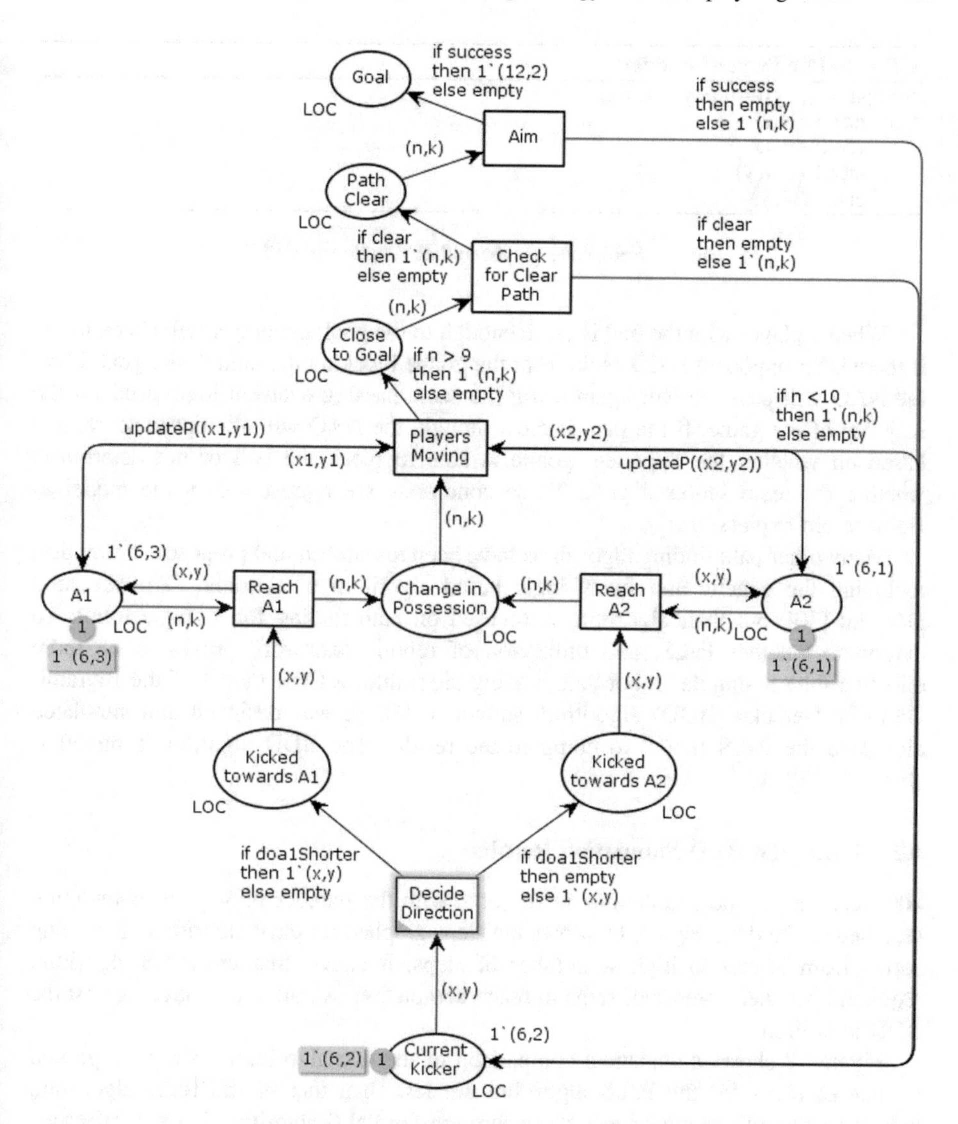

Fig. 7. PaLS modeled as a CPN

When the ball is passed off to another player, the receiving player becomes the player in possession of the ball, and kicker will become one of the other ally players that can be passed to. The model accounts for this by moving the coordinates around to match. The ally NAOs not in possession of the ball will then move forward towards the goal to get into a position where they may receive the ball next. The model uses the self-made update player function, updateP, to move the players forward after the ball has been passed, incrementing the position of the player on the field until they reach the goal area of the field at x = 11 or x = 12. The updateP algorithm is shown in Fig. 8.

CPN Update Player Function

```
fun updateP((x,y)) = if (x,y) = (11,2)
      then 1`(11,2)
      else if x < 12
      then 1`(x+1, y)
      else 1`(11, y)
```

Fig. 8. Self-made player update function in CPNT

When a player with the ball is close enough to the goal, the player will check to see if there is an opponent NAO aside from the goalie blocking the path to the goal. If so, the NAO will pass the ball again using the same passing decision logic used for the majority of the game. If the path is clear though, the NAO will attempt to score, and based on whether the opponent goalie is able to block the ball or not determines whether the team scores a goal. These conditions are represented in the model as Boolean arc expressions.

Many other path finding algorithms have been researched and proposed in robotics, including the genetic and fuzzy logic based algorithm of Bajrami, Dërmaku, and Demaku [30], but their algorithm is focused on path finding for a single robot. To determine whether PaLS, and utilization of robotic teamwork proves to be more effective than a singular robot path finding algorithm, a CPN model of the Bajrami-Dërmaku-Demaku (BDD) algorithm, shown in Fig. 9, was designed and simulated alongside the PaLS model to compare the results. The BDD algorithm's model is shown in Fig. 10.

4.2 PaLS and BDD Simulation Results

200 iterations of the simulations were done, with the number of steps to completion recorded as the data. Figure 11 shows the steps graphed for each algorithm after being sorted from lowest to highest number of steps. It shows that the PaLS algorithm required a smaller number of steps to reach completion overall when pitted against the BDD algorithm.

Figure 12 shows a statistical comparison of the two algorithms. The average and median of steps for the PaLS algorithm are less than that of the BDD algorithm, indicating that utilization of teamwork through the PaLS algorithm is more effective, and will require fewer steps to reach a goal. Based on these simulation results, the PaLS algorithm is overall determined to be 34% more efficient than the BDD algorithm.

4.3 Modeling of iPaLS

PaLS demonstrated that using a team-based strategy for gameplay, similar to one humans would use in soccer, will allow robot teams to be more successful than those that use path optimization based strategies. Though, as mentioned before, there was still much improvement to be made on PaLS. iPaLS takes the CPN model created for PaLS and expands upon it, similar to how the iPaLS algorithm expanded upon the PaLS algorithm. Figure 13 shows the CPN model produced for the iPaLS algorithm.

BDD Algorithm

```
1:   if x_b < 10
2:         moveForward()
3:   else
4:         if (x_b, y_b) = (12,1) OR (x_b, y_b) = (12,3),
5:               then (x_b, y_b) = (11, y_b )
6:         else
7:               if clear = true,
8:                     then attempt goal shot
9:               else
10:                    decideDirection()
11:
12:  function moveForward() {
13:        decideDirection()
14:        choose fitness function for fuzzy algorithm
15:        calculateSC()
16:        y_b = y_b + sc
17:  }
18:  function decideDirection() {
19:        left, middle, or right based on sensors
20:  }
21:  function calculateSC() {
22:        os = optimized steps count
23:        as = actual steps count
24:        sc = steps counter for recalculating direction 25:        toward target
26:        if as < os
27:           os = as;
28:        sc = os;
29:        return sc;
30:  }
```

Fig. 9. Bajrami-Dërmaku-Demaku (BDD) algorithm

In the creation of the Colored Petri net for iPaLS, the opponent strategy for any robot that intercepts the ball is to try and move the ball towards the goal – no passing involved. Since opponent possession of the ball meant that certain players (A2 and "current kicker", A3) need to move in the opposite direction, the Petri net has player A3 attempt to follow the movement of the ball while player A2 moves back 1 x space at a time, following the CPN function shown in Fig. 14.

4.4 iPaLS Simulation Results

For the simulation results for iPaLS, the Petri net was simulated 200 times to observe how many times the allies would score against the opponents and vice versa. This is different from the method of simulation for PaLS, in which a separate CPN was created for comparison. Figures 15 and 16 show the results found.

In the previous work, the two algorithms, PaLS and BDD, were compared using the number of steps to completion, or the number of transitions made through the Petri net until the goal is reached. For iPaLS vs. opponents, who only constantly kick towards the goal, the CPN model simulates a game in the sense that one side has scored when

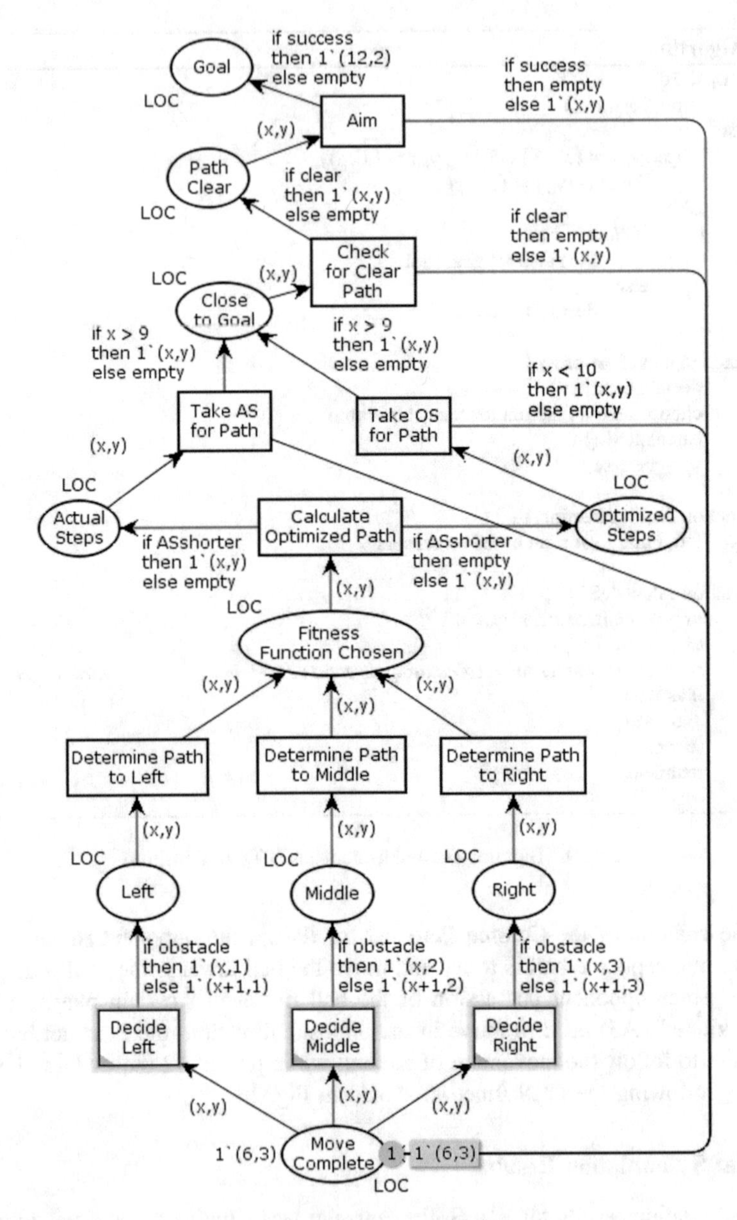

Fig. 10. BDD modeled as a CPN

the simulation has reached completion. The results in the newer model show that using the iPaLS strategy, allies will score a goal over the opponents 82.5% of the time while the opponents will only score 17.5% of the time, as shown in Fig. 15. Figure 16 shows a statistical comparison between the two, comparing the average and median number of

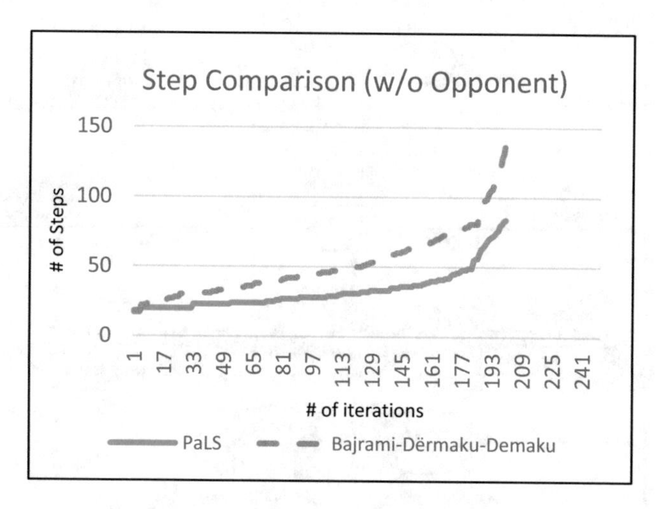

Fig. 11. Step comparison for PaLS and BDD algorithms

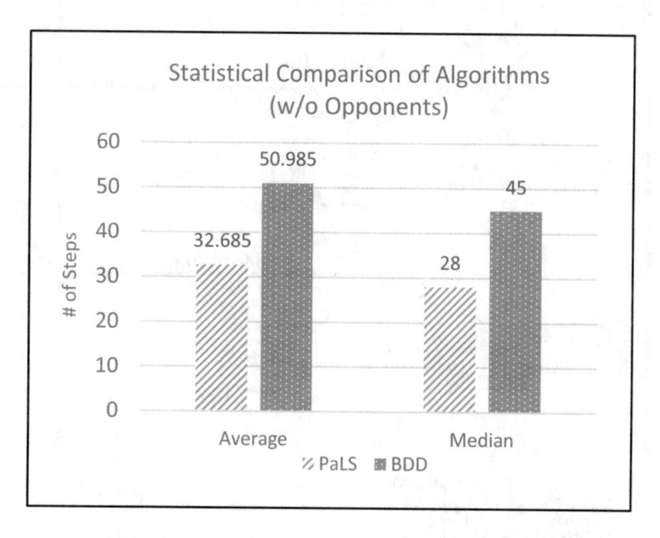

Fig. 12. Statistical comparison graph for PaLS

steps. For both the average and the median, the number of steps was less for the iPaLS algorithm, but with even better results than what was recorded for the PaLS algorithm.

It is important to bear in mind though that the Petri net has limitations in its ability to simulate. Many of the possible actions and paths that the Petri net can take are based on 50/50 chance due to the nature of Boolean variables in CPN Tools, but in reality there are many cases where the paths and decisions that favor the allies would be greater than a 50% chance.

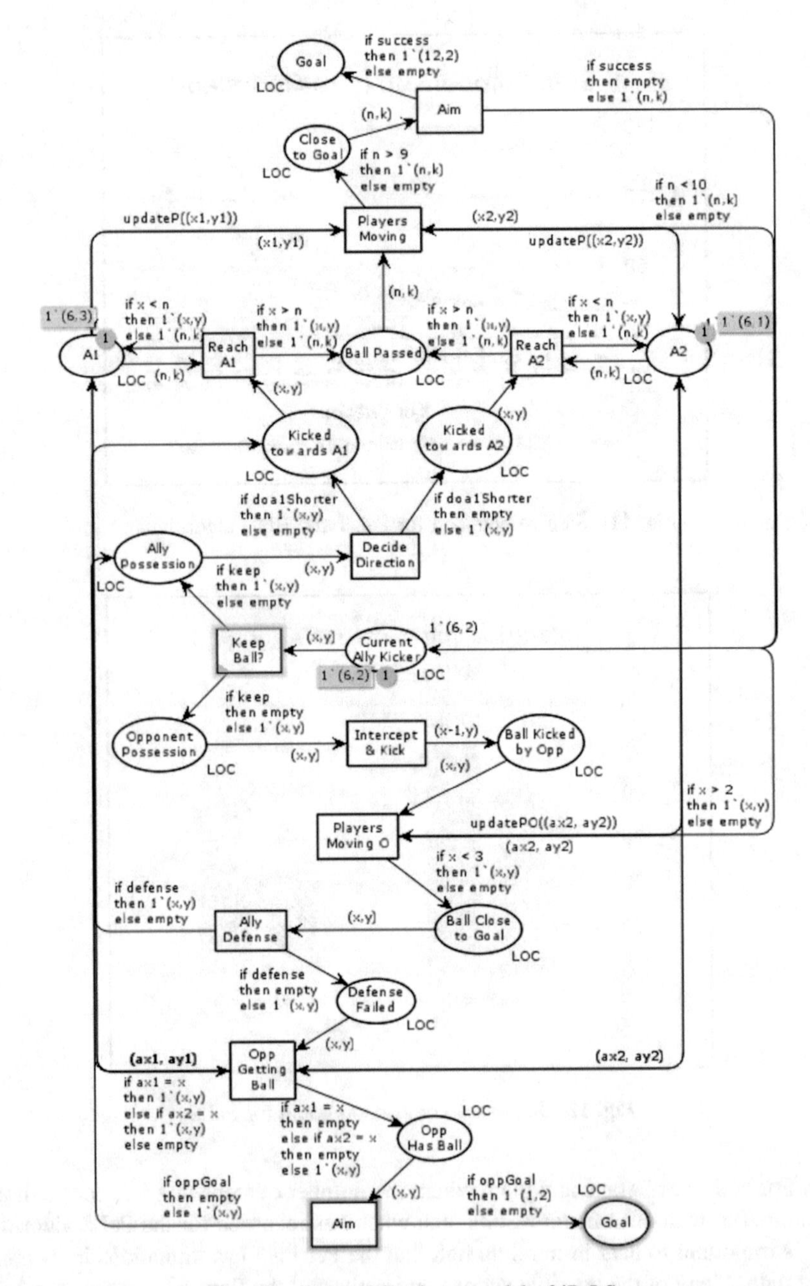

Fig. 13. Colored Petri net model of iPaLS

CPN Update Player during Opposition Function

```
fun updatePO((x,y)) = if (x,y) = (2,y)
    then 1`(2,y)
    else 1`(x-1, y)
```

Fig. 14. Self-made player update during opposition function in CPNT

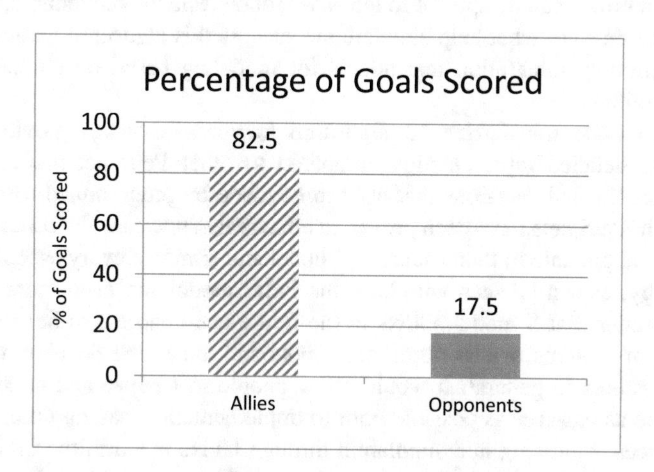

Fig. 15. Allies and opponent iPaLS comparison of goals scored

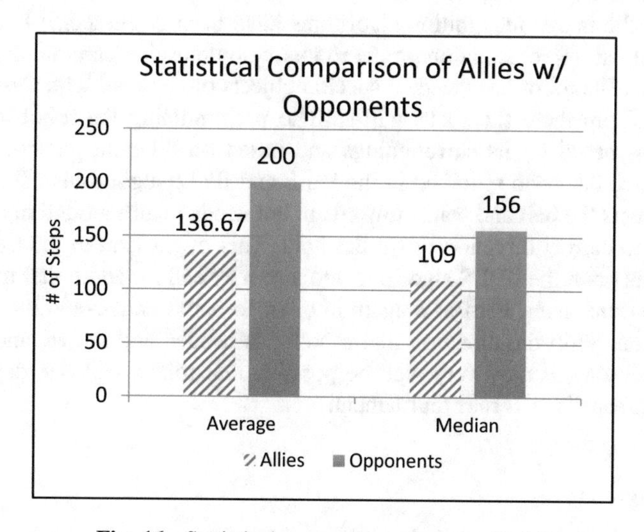

Fig. 16. Statistical comparison graph for iPaLS

5 Conclusions

It is clear that the use of a teamwork strategy such as that of iPaLS is able to more effectively succeed in scoring goals and overcoming the opposing team during a soccer match. With the advancing technologies for image recognition, accurate robot positioning, and ball kicking, RoboCup SPL teams should think more deeply about how the robots can play more so like a human. iPaLS emphasizes the importance of teamwork in soccer and how it can be applied to the NAO robots, and the modeling capabilities of Colored Petri nets are what help simulate the success this algorithm would have, with their programming integration that allows for modeling based on mathematical and logical conditions.

CPNs facilitate transferring an algorithm to a model while providing a more accurate and detailed representation compared to other Petri net models. With the numerous actions and decisions that need to be made by programmed robots during a soccer match, Petri nets have been proven to effectively showcase the concurrent nature of systems and are able to then accurately model and simulate every aspect of a soccer game strategy, as can be seen with both the PaLS model and even more so with the newer and better iPaLS model. CPNs in the future can continue to be used to model other kinds of systems or algorithms for robots and aid in the development of more human-like robots in general. It would allow people to analyze and crosscheck their systems to be as bug-free as possible prior to implementation, saving time, energy, and other resources. Modeling and simulation through CPNs is worth the initial efforts to save added time and stress due to bugs and errors that could be found at a later time, especially for algorithms as complex as those that are needed to program soccer-playing strategies onto NAO robots.

Since the strategy behind the iPaLS algorithm has proven to be effective, and the structure for the image recognition algorithms have been discussed, it is time to look more seriously at implementation on the robots. Simple testing can initially be done to signal whether the robot has detected specific objects on the field with OpenCV image recognition. From there the focus will shift to programming the robot to move and position itself based on its surroundings and based on different gameplay scenarios (some of which has been specified in the PaLS and iPaLS algorithms). For example if the robot detects the ball and walks towards it, but needs to turn around in order to kick the ball in the correct direction down the field. This might lead to modifications and improvements upon the iPaLS algorithm and also may call for additional modeling and testing to be done using Petri nets again in order to better understand how the system might function. With the direction that robotics is headed and the advancements that are made each day, it may very well be possible that robots will one day be able to overcome humans in a soccer tournament.

References

1. Didandeh, A., Mirbakhsh, N., Afsharchi, M.: Concept learning games: an ontological study in multi-agent systems. Inf. Syst. Front. **15**(4), 653–676 (2013)
2. NAO Robot Documentation, Aldebaran. https://www.aldebaran.com/en

3. Alkhalifah, A., Alsalman, B., Alnuhait, D., Meldah, O., Aloud, S., Al-Khalif, H., Al-Otaibi, H.: Using NAO humanoid robot in kindergarten: a proposed system. In: Proceedings of the IEEE International Conference on Advanced Learning Technologies, pp. 166–167 (2015)
4. Alam, M., Vidyaratne, L., Wash, T., Iftekharuddin, K.: Deep SRN for robust object recognition: a case study with NAO humanoid robot. In: Proceedings of the IEEE South East Conference, (SoutheastCon2016), pp. 1–7 (2016)
5. RoboCup International, RoboCup Standard Platform League. http://www.tzi.de/spl/bin/view/Website/WebHome
6. Tang, Y., Cerutti, F., Oren, N., Bisdikian, C.: Reasoning about the impacts of information sharing. Inf. Syst. Front. **17**(4), 725–742 (2015)
7. Albani, D., Youssef, A., Suriani, V., Nardi, D., Dloisi, D.: A deep learning approach for object recognition with NAO soccer robots. In: RoboCup 2016: Robot World Cup XX, pp. 392–403. Springer, Heidelberg (2016)
8. Qian, Y., Baucom, A., Han, Q., Small, A., Buckman, D., Tian, Z., Lee, D.: The UPennalizers RoboCup standard platform league team description paper 2016. Technical paper (2016). https://fling.seas.upenn.edu/~robocup/files/2016Report.pdf
9. Ashar, J., Ashmore, J., Hall, B., Harris, S., Hengst, B., Liu, R., Zijie, M., Pagnucco, M., Roy, R., Sammut, C., Sushkov, O., The, B., Tsekouras, L.: RoboCup SPL 2014 Champion Team Paper, RoboCup 2014: Robot World Cup XVIII, pp. 70–81. Springer, Heidelberg (2014)
10. MacAlpine, P., Depinet, M., Liang, J., Stone, P.: UT Austin villa 2015: RoboCup 2014 3D simulation league competition and technical challenges champions. In: Proceedings of the RoboCup International Symposium 2015 (RoboCup 2015), pp. 118–131 (2015)
11. Chen, S., Ke, J., Chang, J.: Knowledge representation using fuzzy Petri nets. IEEE Trans. Knowl. Data Eng. **2**(3), 311–319 (1990)
12. Kim, S.-y.: Modeling and analysis of a web-based collaborative information system: petri net-based collaborative enterprise. Int. J. Inf. Decis. Sci. **1**(3), 238–264 (2009)
13. Kuo, C., Lin, I.: Modeling and control of autonomous soccer robots using distributed agent oriented Petri nets. In: Proceedings of the IEEE International Conference on Systems, Man, and Cybernetics, pp. 4090–4095 (2006)
14. Zouaghi, L., Alexopoulos, A., Wagner, A., Badreddin, E.: Mission-based online generation of probabilistic monitoring models for mobile robot navigation using Petri nets. Robot. Auton. Syst. **62**, 61–67 (2014)
15. Kim, S.-y., Yang, Y.: A self-navigating robot using fuzzy Petri nets. Robot. Auton. Syst. **101**, 153–165 (2018)
16. Jensen, K.: Coloured Petri nets. In: IEE Colloquium on Discrete Event Systems: A New Challenge for Intelligent Control Systems, London, pp. 5/1–5/3 (1993)
17. Jensen, K., Kristensen, L.: Colored Petri nets: a graphical language for formal modeling and validation of concurrent systems. Commun. ACM **58**, 61–70 (2015)
18. Bonilla, B., Asada, H.: A robot on the shoulder: coordinated human-wearable robot control using Coloured Petri nets and Partial Least Squares predictions. In: Proceedings of IEEE International Conference on Robotics and Automation (ICRA 2014), pp. 119–125 (2014)
19. Farinelli, A., Marchi, N., Raeissi, M., Brooks, N., Scerri, P.: A mechanism for smoothly handling human interrupts in team oriented plans. In: Proceedings of the 2015 International Conference on Autonomous Agents and Multiagent Systems, (AAMAS 2015), pp. 377–385 (2015)
20. Murata, T.: Petri nets: properties, analysis and applications. Proc. IEEE **77**, 541–574 (1989)
21. CPN tools, Colored Petri net tools. http://cpntools.org
22. Khandelwal, P., Hausknecht, M., Lee, J., Tian, A., Stone, P.: Vision calibration and processing on a humanoid soccer robot. In: Proceedings of the Fifth Workshop on Humanoid Soccer Robots, Nashville, pp. 71–76 (2010)

23. Härtl, A., Visser, U., Röfer, T.: Robust and efficient object recognition for a humanoid soccer robot. In: RoboCup 2013: Robot World Cup XVII, pp. 396–407 (2013)
24. Naushad Ali, M., Abdullah-Al-Wadud, M., Lee, S.: An efficient algorithm for detection of soccer ball and players. In: Proceedings of Conference on Signal Processing Image Processing (SIP 2012), pp. 1–8 (2012)
25. Neves, A., Trifan, A., Dias, P., Azevedo, J.: Detection of aerial balls in robotic soccer using a mixture of color and depth information. In: Proceedings of the IEEE International Conference on Autonomous Robot Systems and Competitions, pp. 227–232 (2015)
26. Cheng, Q., Yu, S., Yu, Q., Xiao, J.: Real-time object segmentation for soccer robots based on depth images. In: Proceedings of the IEEE International Conference on Information and Automation (ICIA 2016), pp. 1532–1537 (2016)
27. Mulya, A., Ardilla, F., Pramadihanto, D.: Ball tracking and goal detection for middle size soccer robot using omnidirectional camera. In: Proceedings of the International Electronics Symposium (IES 2016), pp. 432–437 (2016)
28. Pulli, K., Baksheev, A., Kornyakov, K., Eruhimov, V.: Real-time computer vision with OpenCV. Commun. ACM **55**(6), 61–69 (2012)
29. Pham, T., Cantone, C., Kim, S.-Y.: Colored Petri net representation of logical and decisive passing algorithm for humanoid soccer robots. In: Proceedings of the IEEE International Conference on Information Reuse and Integration, pp. 263–269 (2017)
30. Bajrami, X., Dërmaku, A., Demaku, N., Maloku, S., Kikaj, A., Kokaj, A.: Genetic and fuzzy logic algorithms for robot path finding. In: Proceedings of the 5th Mediterranean Conference on Embedded Computing (MECO), Bar, pp. 195–199 (2016)

Analyzing Cleaning Robots Using Probabilistic Model Checking

Rafael Araújo[1(✉)], Alexandre Mota[1], and Sidney Nogueira[2(✉)]

[1] Centro de Informática, UFPE, Av. Jornalista Anibal Fernandes, Cidade Universitária, s/n, Recife, PE 50740-560, Brazil
{rpa4,acm}@cin.ufpe.br

[2] Departamento de Computação, UFRPE, R. Dom Manoel de Medeiros, s/n, Dois Irmãos, Recife, PE 52171-050, Brazil
sidney.nogueira@ufrpe.br

Abstract. Robots are becoming ubiquitous, even in our homes. Nevertheless simulations and filming their behavior still are the most frequently used techniques to analyze them. This can be inefficient and dangerous. This work considers Probabilistic Model Checking (PMC) to perform such an analysis. With PMC we can verify whether a robot trajectory (described in terms of an algorithm) satisfies specific behaviors or properties (stated as temporal formulas). For instance, we can measure energy consumption, time to complete missions, etc. As consequence we can also determine if an algorithm is superior to another in terms of such properties. We choose the PRISM language; it can be used in more than one PMC tool. We also propose a DSL to hide the Prism language from the end user; this DSL passes an automatic sanity test. Programs and properties written in the proposed DSL are automatically translated into the notation of PRISM by following a set of mapping rules. An integrated environment is proposed to support the authoring of the algorithms and properties in the proposed DSL and checking them in a Probabilistic Model Checker. Finally, we present an evaluation on three algorithms towards our proposal.

1 Introduction

Robots can replace heavy, difficult, and dangerous tasks in industry as well as in house cleaning. Therefore they are becoming ubiquitous, specially in industry due to save costs [24].

However, robots need energy to work and a plan (behavior) to follow. Robot builders generally use simulations (recording robots motion during a certain time [27]) to check robot behaviors, energy consumption, and other aspects as well. Thus in general robots cannot use the best solution in terms of energy consumption nor provide the safest behavior. Naturally industry tries to use robots in very specific problems so that an optimal algorithm (aided by many sensors as it is financially feasible) can possibly be found for some specific task, for example,

T. Bouabana-Tebibel et al. (Eds.): IEEE IRI 2017, AISC 838, pp. 23–51, 2019.
https://doi.org/10.1007/978-3-319-98056-0_2

to clean some region. For instance, Roomba[1] and Neato[2] are vacuum-cleaners. Roomba is a kind of blind robot when compared to Neato. Roomba can only detect nearby obstacles. Neato can do the same as Roomba but also uses a laser to map the space to clean before starting its work and thus it can possibly take less time and spend less energy to do the same job. But Neato is more expensive than Roomba because extra hardware is employed.

This work uses a different strategy. We use probabilistic model checking [8] (PMC). With PMC, we can go beyond simulation and verify desirable properties using temporal logic. We can also compare motion algorithms by numerical measures, determine whether the mission (for instance, cleaning all parts) can be completed, how long it will take, how much energy it will need, etc. (in the direction of the work [30]). PMC is formulated in terms of Markov chains, which have several applications in Engineering [16].

To concretely illustrate our proposal we consider the iRobot Create 2 [1][3] programmable vacuum cleaning robots as case study. Its factory behavior is very simple: When it reaches an obstacle, it changes its path in some way (the specific algorithm is an industrial property and here we propose simple alternatives to illustrate its behavior). Some more advanced models can recharge themselves automatically when necessary, but we do not consider such a feature in our analysis. Instead we check whether the robot has enough energy to complete its mission.

To ease the use of our work, we provide a domain-specific language (DSL), whose semantics is given in terms of a PMC language via a series of mapping rules. This DSL can describe the robot motion algorithm, environment and properties of interest. Moreover, we perform a kind of sanity test on our DSL. From the DSL grammar, several valid DSL scenarios are automatically created with the LGen tool [10]. We translate each of these valid DSL scenarios in terms of the PMC language and check whether they are compilable. Such a verification assures we cannot create invalid PMC specifications.

As result, the main contributions of this work are:

- A discrete and abstract characterization of an iRobot equipment as well as its environment stated in PRISM;
- A domain-specific language to describe motion planning algorithms and their analyses;
- Translation rules to map the DSL (motion planning algorithms and properties related to check mission completion, measure number of steps the robot takes, energy consumption, and time to complete a mission) into the Prism language;
- An integrated environment to allow writing algorithms and properties in the proposed DSL and checking them in a Probabilistic Model Checker.

This work is organized as follows. In Sect. 2 we present the robot and environment we are focusing in this work. Section 3 presents our abstraction of such

[1] http://www.irobot.com/For-the-Home/Vacuuming/Roomba.aspx.
[2] https://www.neatorobotics.com/robot-vacuum/botvac/.
[3] http://www.irobot.com/About-iRobot/STEM/Create-2.aspx.

a scenario. In Sect. 4 we present a DSL to avoid using PRISM directly and an environment created for supporting the authoring of algorithms. Sect. 5 considers a straightforward sanity test for this DSL. At this point, we perform an evaluation of our proposed DSL in Sect. 6. Related work is discussed in Sect. 7. And finally, we present our main conclusions and future work in Sect. 8.

2 An iRobot Usage Scenario

In this work we focus on a domestic robot: the iRobot Roomba. It is a cleaning robot that works in rooms with walls and obstacles as a chairs and cabinets. Fig. 1 illustrates the robot in a real environment.

Fig. 1. iRobot in a room

Each Roomba robot comes with a different embedded (proprietary) algorithm, however the basic robot behavior consists in:

- Traverse the room in some direction until an obstacle is reached; when the sensors detect an obstacle (or the room boundaries) the robot takes another direction;
- Spin in the middle of the room, even without reaching an obstacle;
- When the battery level is low, some versions simply stop working whereas others auto recharge going back to recharging basis.

As such behaviors are very complex and the industry, in general, does not use any formal models to analyze such robots, a frequent technique to determine whether an algorithm is better than another is to film the robot behavior and run simulations [24,27]. Fig. 2 illustrates the robot filming[4]. In this figure, one can see the robot trajectories that follow the patterns listed above.

[4] https://dornob.com/time-lapse-photo-reveals-robotic-vacuum-paths/.

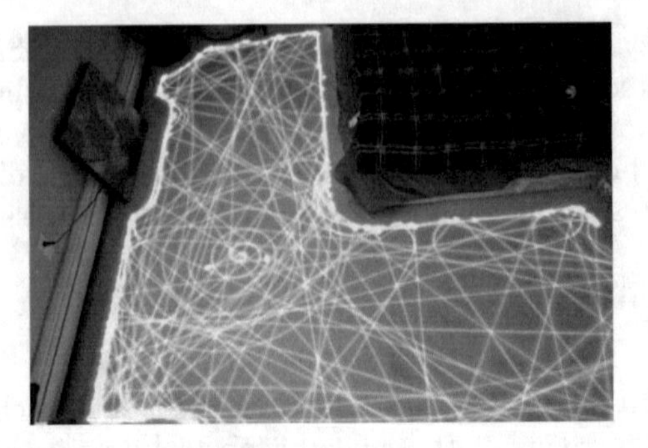

Fig. 2. Filming of Robot's behavior

3 Abstracting the Real-World Scenario

In this section we abstract the robot and its environment in terms of a Mathematical model. Fig. 3 is an abstract representation of the elements in Fig. 1: the room is represented as a matrix and the robot and obstacles elements of the matrix. Formally, the abstract model is formed of different modules that work together to represent the robot behavior inside a particular environment: the robot module, a matrix module and the obstacle modules.

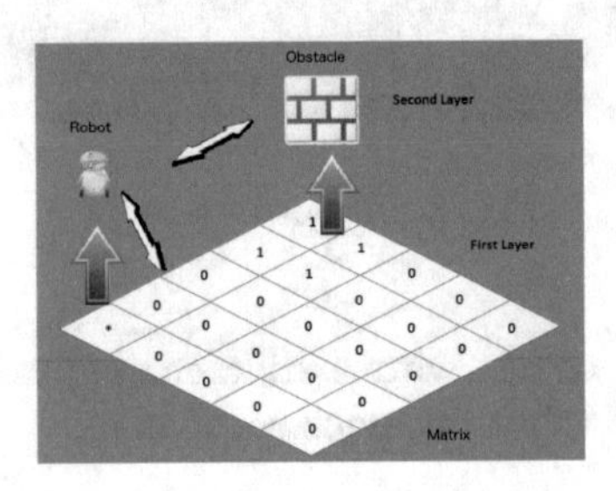

Fig. 3. Model's layers and elements

From Fig. 3, we can identify three main entities (or modules):

1. **Environment.** It is represented as a *matrix*, where 0's are navigable cells and 1's are inaccessible cells, as presented in Fig. 3. Each cell has a state that is updated when the robot cleans (passes over) the cell. This state is necessary to measure the robot cleaning coverage;
2. **Robot**. The *robot*, visually illustrated by an asterisk in the environment (matrix), is modeled as a separate module. The relevant robot state is its position (in the matrix), commands to be executed and its program counter. We consider the robot size equals the size of a single cell;
3. **Obstacles**. Obstacles are rectangle-based elements, visually illustrated by 1's in the environment (matrix).

The main idea behind this abstract model is to describe the elements in terms of a formal specification language (PRISM) introduced in Sects. 4.2 and 4.3, and then show how a user-friendly DSL language described in Sect. 4.1 can be translate into this formal language.

4 A DSL for Robot Motion and Analysis

To hide the PRISM model for the robot, and to give a more user-friendly experience, we propose a Domain-Specific Language (DSL) for supporting the robot motion specification and analysis. The DSL has the following blocks:

- `Prog`: the robot motion specification and the properties to be analysed;
- `Init`: environment dimensions and robot start position;
- `Properties`: the properties to be analysed;
- `Behavior`: the algorithmic description of the robot's behavior.

Below we can find a valid (and very simple) DSL example named `robot`:

```
prog robot

    init
        Map(5, 5)
        Position(2, 1)
    endinit

    properties
        Terminates
        Energy
        Time
    endproperties

    behavior
        if (downIsClear()) {
            Down()
        }

        if (rightIsClear()) {
            Right()
        }

        if (downIsClear()) {
            Down()
        }

        while (leftIsClear()) {
            Left()
        }
    endbehavior

endprog
```

The `robot`'s initialization section states that a 5×5 matrix is its environment (the matrix content comes with an external ASCII as shown in the sequel); and, the robot is initially located at position (2,1), namely second column and first line. Its behavior is an attempt to go downward one cell, then to the right one cell, go downward again one, and finally try to go to the left as much as possible. The `robot`'s properties specifies we are interested in three properties to be analyzed: termination, energy and time. This means that we want to know if the robot can finish its mission[5] (`Terminates`), how much energy it will need (`Energy`), and how much time it needs to accomplish the mission (`Time`). Finally, in the behavior section we have the robot algorithmic motion planning.

As explained previously, the matrix specification comes in an external file. For this example, the matrix is the 5×5 matrix that follows. In such matrix, 0's stand for free cells and 1's obstacles.

```
0,0,0,1,1
0,0,1,1,1
0,1,1,1,1
1,1,1,1,1
1,1,1,1,1
```

4.1 Syntax of the DSL

Our top-level grammar element is a `Program` (boldface font is used for tokens and italic stands for terminal/non-terminal productions) stated as follows.

```
[Program] ⟹ prog ID
               Init
               Properties
               Behavior
             endprog
```

As we saw before, the scope of a `Program` is determined by the tokens **prog** and **endprog**, whose name is `ID`. It has three sections: `Init`, `Properties`, and `Behavior`.

Its initialisation section has two non-terminals as follows.

```
[Init] ⟹ init
            MapDefinition
            Position
          endinit
```

The first one, `MapDefinition`, is expressed by.

```
[MapDefinition] ⟹ Map(INT, INT)
```

where `INT` means a natural number to state the size of the map (keyword **Map**).

The `Position` element states the robot's starting position[6] as follows.

```
[Position] ⟹ Position(INT, INT)
```

The next section is `Properties`, described as follows.

```
[Properties] ⟹ properties
                   Property+
                 endproperties
```

[5] Finish a mission means the robot can pass over all free cells of the matrix at least once.

[6] Our tool checks whether the starting position lies in the map's dimensions.

The `Property` element represents a property to be analyzed; ($Property^{+}$) one or more properties. As already introduced, there are three different properties.

```
[Property] ⟹   Terminates
           | Energy
           | Time
```

Our last section is `Behavior`, where the robot's motion algorithm takes place.

```
[Behavior] ⟹ behavior BehaviorInstruction+ endbehavior
```

The algorithm has one or more `BehaviorInstruction` from the following options.

```
[BehaviorInstruction] ⟹   StandaloneCommand
                      | Choice
                      | ControlStructure
```

The `StandaloneCommand` element comprehends one of the following commands.

```
[StandaloneCommand] ⟹   Up()
                    | Down()
                    | Left()
                    | Right()
```

Each of the above commands move the robot to the adjacent cell in the direction indicated by the command, if the adjacent cell is inside the boundaries of the matrix, and there is no obstacle in the destination cell. Otherwise, the robot position remains the same. For instance, the command **Up** moves the robot to the adjacent cell northwards.

The `Choice` instruction represents a (optional probabilistic) choice between two possible behaviors. The element has two constructors, denoted as:

```
[Choice] ⟹   choice(StandaloneCommand, StandaloneCommand)
         | choice(REAL, StandaloneCommand, StandaloneCommand)
```

Note that the second **choice**(.) constructor uses a real number. Let **choice**(p, C_1, C_2) be such a probabilistic choice. Then the chance of C_1 to occur is p and that for C_2 is 1 - p. The first constructor sets 50-50% chance for each command.

The `ControlStructure` element represents one of the control structures that follow

```
[ControlStructure] ⟹   If
                   | IfElse
                   | While
```

For simple branches, we have

```
[If] ⟹ if (Evaluation) {
    BehaviorInstruction+
  }
```

and

```
[IfElse] ⟹ if (Evaluation) {
    BehaviorInstruction+
  } else {
        BehaviorInstruction+
  }
```

For repetition we have

```
[While] ⟹ while (Evaluation) {
        BehaviorInstruction+
        }
```

In all control structures, `Evaluation` is one of the following options.

```
[Evaluation] ⟹    upIsClear()
              | downIsClear()
              | leftIsClear()
              | rightIsClear()
```

The boolean function **upIsClear** checks whether the robot can move upward. The other functions work accordingly to their name.

4.2 Probabilistic Model Checking

Probabilistic model checking [3] is a complementary form of model checking aiming at analyzing *stochastic* systems. The specification describes the behavior of the system in terms of rates (or probabilities) in which a transition can occur.

Probabilistic model checkers can be used to analyze quantitative properties of (non-deterministic) probabilistic systems by applying rigorous mathematics-based techniques to establish the correctness of a certain property. The use of the probabilistic model checkers reduces the costs during the construction of a real system by verifying in advance that a specific property does not conform to what is expected about it. This is useful to redesign models.

There are some tools that specialize in probabilistic model checking. The most well-known are: PRISM [18], Storm [7], PEPA [26], and MRMC [15].

This work focuses in the syntax of the language PRISM, which can be analyzed by the PRISM tool, the Storm model checker and other probabilistic model checkers as well. The next section gives an overview of PRISM.

The PRISM Language. The PRISM language [18] is a probabilistic specification language designed to model and analyze systems of several application domains, such as multimedia protocols, randomized distributed algorithms, security protocols, and many others.

The PRISM tool uses a specification language also called *PRISM*. It is an ASCII representation of a Markov chain/process, having states, guarded commands and probabilistic temporal logics such as PCTL, CSL, LTL and PCTL.

PRISM can be used to effectively analyze probabilistic models such as:

- Discrete-Time Markov Chains (DTMCs)
- Continuous-Time Markov Chains (CTMCs)
- Markov Decision Processes (MDPs)
- Probabilistic Automata (PAs)
- Probabilistic Timed Automata (PTAs)

As the above models use different analysis algorithms, we need to tell PRISM the model type we are interested in the beginning of a PRISM specification as shown below:

```
<type of probabilistic model>

<definitions>

module <name>
(...)
endmodule
```

where `<type of probabilistic model>` can be one of the keywords: *ctmc, dtmc, mdp*, etc. In the `<definitions>` section, we state constants and formulas. These can be used to build the model as well as analyze it. The `<name>` must be a valid and unique identifier in the whole specification. It can be used to compose with other modules. The existence of more than one module implies the parallel execution of the modules. Guards on shared state variables control state changes.

To describe the transitions in the model, a command-based syntax is used:

```
[] guard -> prob_1 : update_1 + ... + prob_n : update_n;
```

where `guard` is a condition (a composite proposition formed by conjunctions (&), disjunctions (|) or negation (!)) that enables the probabilistic transitions, such that `prob_1`,..., `prob_n` are the probabilities (or rates) associated with the transitions and the `update_1`,..., `update_n` are commands describing eventual state changes.

In the PRISM notation, comparisons (evaluations) are expressed by a single equals (=) sign. For instance, the boolean expression `(state = 10)` evaluates to true if the variable `state` equals `10`. Moreover, the notation for binding a value to a variable is `variable_name'= value`, which indicates a state change. As an example, the PRISM command `(c'= c+1)` expresses the variable `c` is incremented.

We can analyze quantitative properties of a PRISM specification by submitting probabilistic temporal logic formulas to the PRISM tool. Formally, given a desired property, expressed as a temporal logic formula f, and a model M with initial state s, decide whether $M,\ s\ \models\ f$. In PRISM this is encoded as:

```
P = ? [ f ]
```

The term `P = ?` means "what is the probability of formula f be valid?".

It is usual that `f` mentions specific states of the model. But eventually it is interesting to abstract these states in terms of *labels*. Those *labels* contain a state-related expression that results a boolean value (*true* or *false*) when evaluated.

Example of a property with a label is:

```
P = 1 [ F "terminate" ]
```

with "`terminate`" being a label for an expression, such as `(state = 10)`. This command asks PRISM to check whether eventually (`F`) the variable `state` reaches value `10`, abstracted as `terminate`, with a probability of 100%.

PRISM supports the specification and analysis of properties based on rewards, which allow values to be associated with certain states or transitions of the model. This enables PRISM to perform quantitative analysis of properties in addition to the analysis of probabilities. Rewards are used in this work to model and measure, among other metrics, expected energy consumption for a robot. For example, the following PRISM fragment

```
rewards "energy"
    true : 1;
endrewards
```

defines a reward named "`energy`" that assigns value `1` to every state of the model. In general, the left-hand part of a reward (`true` in the fragment) is a guard, and the right-hand part is a reward (`1` in the fragment). States of the model which satisfy the predicate in the guard are assigned the corresponding reward (cumulative value). PRISM allows the analysis of rewards. For instance,

```
R={"energy"} =? [ F counter = total ]
```

checks the value for the `energy` rewards when the counter variable equals the value of the `total` constant.

Numerical vs Statistical Model Checking. Numerical analysis based on uniformisation and statistical techniques based on sampling and simulation are two distinct approaches for transient analysis of stochastic systems [31]. PRISM can perform an approximate analysis by means of a technique often called *Statistical Model Checking*; it relies on highly efficient sequential acceptance sampling tests that enables statistical model checking to be performed very efficiently, without explicitly constructing the corresponding probabilistic model. Thus, statistical model checking is particularly useful on very large models when normal model checking is infeasible.

To use Statistical Model Checking in PRISM, one needs to use Continuous-time Markov chains (CTMCs). Therefore, a problem naturally suited to Discrete-time Markov chain (DTMC) description can be modeled as a Continuous-time Markov chains (CTMCs) to benefit from the efficient power provided by Statistical Model Checking. This is the main reason in this work the abstract model for the robot and its environment is stated in terms of CTMC.

4.3 Modelling and Analysing the Abstract Scenario in PRISM

In this section we present PRISM fragments of the work [2], which is based on the work [30], for the three entities (modules) introduced in Section 3 namely: the environment, obstacles, and the robot.

The *Obstacle* Modules. *Obstacles* represent regions that are inaccessible by the robot. In this work we model obstacles as rectangles. Thus we need a point (position (x, y)) and its height (h) and width (w). With these elements, we can create a rectangle (x, y)—(x+w, y+h). In practice, however, PRISM cannot express this representation directly. Instead, we have to break down a rectangle as a series of rows. In what follows we present just a single example of such a row module.

```
module obstacle_1
    obstacle_1_vertical_alignment : [1..n] init 1;
    obstacle_1_left_bound : [1..n] init 6;
    obstacle_1_right_bound : [1..n] init 8;
endmodule
```

The variable **obstacle_1_vertical_alignment** states the position in the y-axis (in this case 1). The variable **obstacle_1_left_bound** denotes the initial position in the x-axis (in this case 6) and the variable **obstacle_1_right_bound** captures the end position in the x-axis (in this case 8). This single module creates the row (6, 1)—(8, 1). By replicating this module and changing the initial position in the y-axis, we get a rectangle.

The *Robot*'s Module. The `robot` module represents the robot motion across the environment. An excerpt is provided below.

```
module robot
    x : [1..n] init 1;
    y : [1..n] init 1;
    movement: bool init true;
    moving: bool init false;
    finished: bool init false;
    instruction_0_repetition_instructions: [0..1] init 0;
    instruction_1_conditional_instructions: [0..1] init 0;
    (...) // rest of the variables, if applicable.
    // Movement reset guard
    [] (movement=true) -> (moving'=true) & (movement'=false);
    (...) // Movement transitions...
    [] (moving=false & movement=false & passable_tiles_completed =
        passable_tiles_number & finished=false & instruction =
        total_instructions)-> processes : (finished'=true);
    [] (moving=false & movement=false & passable_tiles_completed =
        passable_tiles_number & finished=true & instruction=
        total_instructions) -> processes : (finished' = true);
endmodule
```

In the above PRISM excerpt, the variables `x` and `y` capture the robot's initial position in the matrix (position (1, 1)). The variable **movement** indicates if the robot is ready to move, the variable **moving** indicates if the robot is currently moving, and the variable **finished** captures the conclusion of a mission. The variables **instruction_n_repetition_instructions** and **instruction_n_conditional_instructions** are associated to the DSL and thus will be better explained in Sect. 4.4.

The first PRISM transition (`[]`) refers to resetting the robot to its initial state before a next movement (comment "**Movement reset guard**" above). The specific movements are not shown above (comment "**Movement transitions...**"), but in what follows. The last two transitions signals the completion of a mission.

From Fig. 4, we have two situations: a free movement of the robot (Fig. 4(a)) and a movement due to an obstacle (Fig. 4(b)). In both situations, some tasks are performed: the task "**Verify if movement is enabled**", the task "**Check bounds/obstacles**", the task "**Move**", etc. For conciseness, we will just present and explain the task "**Verify if movement is enabled**".

In what follows we consider a PRISM transition divided in three parts as in Fig. 4.

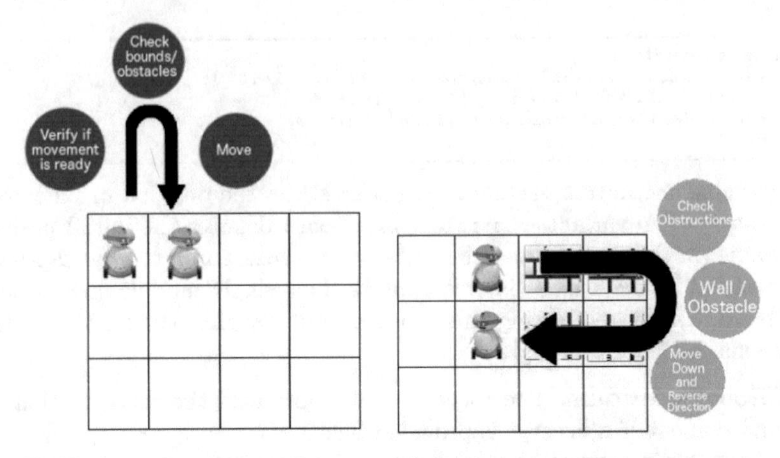

(a) Robot's evaluations in order to move

(b) Robot's evaluations when an obstacle is encountered

Fig. 4. States of the robot movement

Verify if movement is enabled. The "`Verify if movement is enabled`" task is represented by the following (PRISM) proposition.

```
(movement = false) & (moving = false) & (instruction = n)
```

That is, the robot must be stopped and the current instruction (or state) must be `n`.

Check bounds/obstacles. Checking for obstacles or bounds becomes.

```
(((y = obstacle_1_vertical_alignment) & (x + 1 < obstacle_1_left_bound
    | x + 1 > obstacle_1_right_bound)) | (y !=
    obstacle_1_vertical_alignment)) & (x < n)
```

That is, if the robot is at the same vertical position of an obstacle (the y-axis) (`y = obstacle_1_vertical_alignment`), but in a different horizontal position (the x-axis) ($\texttt{x}+1 < \texttt{obstacle\_1\_left\_bound}|\texttt{x}+1 > \texttt{obstacle\_1\_right\_bound}$), or in a different vertical position (`y != obstacle_1_vertical_alignment`). The additional condition check ($\texttt{x} < \texttt{n}$) is an extra boundary check, as the robot can already be adjacent to a wall in the matrix, and the actual position was not evaluated until this condition.

Move. The last part concerns the update in variables after the previous propositions (as the comment `(* previous propositions *)`) were satisfied.

```
[] (* previous propositions *) -> processes : (x' = x + 1) & (movement
    ' = true) & (* command-specific update *);
```

In the above case, we update the horizontal position of the robot `x' = x + 1` (an horizontal move), state the robot is moving and perform specific update commands, omitted here to save space. The interested reader can find them in [2].

***Matrix* Module.** Recall from Sect. 3 that, due to PRISM restrictions, our matrix is vectorised as several row modules. Matrix cells have enumerable states as shown in Table 1.

From Table 1, state 0 means the cell is free to be passed over and state 1 represents an obstacle. State 2 represents a cell that has been passed over once, and the state 3 represents a cell that has been passed over more than once.

Table 1. Matrix's cell states

States	Description
0	Not passed
1	Obstacle
2	Activated cell
3	Passed after activation

The next snippet represents the first row of a matrix. Its interactions with the previous presented modules are omitted to save space but can be found in [2].

```
module matrix_line_1
    line_1_position_y  : [1..n] init 1;
    line_1_column_1    : [0..3] init 0;
    line_1_column_2    : [0..3] init 0;
    line_1_column_3    : [0..3] init 0;
    line_1_column_4    : [0..3] init 1;
    line_1_column_5    : [0..3] init 1;
    (...) // Cell states transitions
endmodule
```

The `line_1_position_y` represents the vertical position of this row. This is used by the robot module, for instance. Each `line_x_column_y` variable contains the actual state of the cell according to Table 1.

For each `line_x_column_y` variable, there are three similar groups of transitions to capture the possible changes in the states of cells.

4.4 A DSL Semantics in PRISM

In this section we give a semantics, in terms of what was explained in Sect. 4.3, to our proposed DSL. Such a semantics is given by applying a set of translation rules following the application ordering shown in Fig. 5. To save space, we just present a subset of the rules. To see all the rules, please refer to [2].

As presented in Fig. 5, the translation starts with Rule 1 (depicted below). The top part of a rule matches the DSL syntax and the bottom part the respective PRISM semantics. Particularly, the top part of Rule 1 matches the non terminals of the program. This rule outputs the main PRISM module (ID), shown in the bottom part of the rule. In such a rule, if the map has the expected

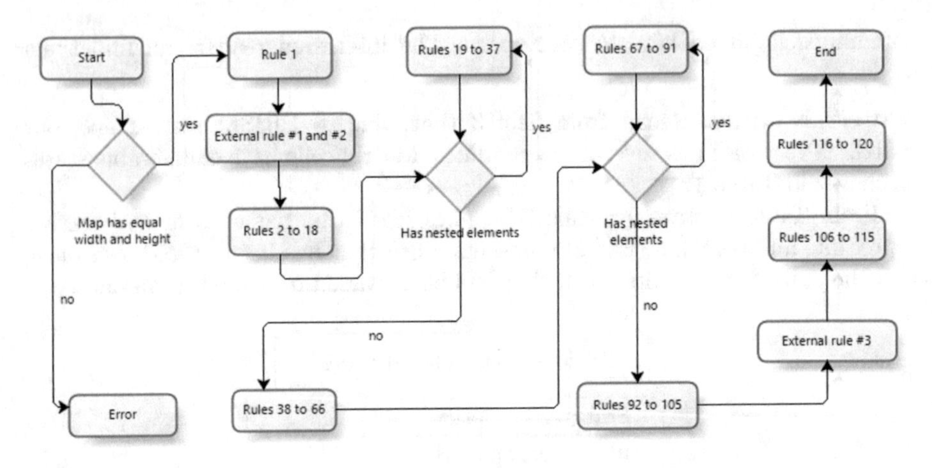

Fig. 5. Rule application flow

dimensions the rule produces the respective output, and subsequent rules are triggered. Otherwise, an error is reported.

Rule 1 (create-model) [[**prog** ID
init *MapDefinition Position* **endinit**
properties *Property+* **endproperties**
behavior *BehaviorInstruction+* **endbehavior**
endprog]] $^{\text{create-model}}$

$\Downarrow$

ctmc
const int n = [[MapDefinition]] $^{\text{generate-size}}$;
...
global moving : bool init true;

module ID
[[Position [[MapDefinition]] $^{\text{generate-size}}$]] $^{\text{generate-robot-coordinates}}$
movement: bool init false;
finished: bool init false;

[[BehaviorInstruction, 0]] ^{iterator}
// Movement reset (static text) guard comes here
[[BehaviorInstruction, 0]] $^{\text{instruction-controller}}$
// (Static text) Transitions related to the mission's end comes here
...
endmodule

[[[[MapDefinition]] $^{\text{generate-size}}$]] $^{\text{generate-obstacles}}$
[[MapDefinition]] $^{\text{generate-matrix-base}}$
[[Property+]] $^{\text{generate-rewards}}$

Note that the **module** `ID` has some local variables as well as transitions concerning movement control, according to what has been presented in Sect. 4.3. Outside this module, other functions based on the map definition (`MapDefinition`) and properties (`Property+`) are called to create the other PRISM elements (modules and formulas).

Auxiliary Variables Generation. To translate conditional and repetition commands, control flow variables (playing a role of a program counter) are declared. For example, the choice uses instruction variables as follows.

> **Rule 5** (instance) [[choice(p1, c1, c2), counter]] ^{instance}
>
> ⇓
>
> **instruction_** [[counter]] **_choice_instructions : [0..2] init 0;**

where `counter` is our internal program counter.

For the `If` control structure, the translation is a bit more elaborate.

> **Rule 9** (instance) [[**if** (evaluation) **{** body **}**, counter]] ^{instance}
>
> ⇓
>
> **instruction_** [[counter]] **_conditional_instructions : [0..** [[body]] ^{length}**] init 0;**
> [[body, (**instruction_** [[counter]] **_sub_instruction_**), 1]] $^{\text{sub-instance}}$

Note that now the upper bound value for the variable `instruction_counter_conditional_instructions` is defined as the conditional's body length and we have sub-instructions to control the conditional's internal program counter.

Other control variables are introduced by the respective rules.

Robot's Coordinates. The following rule creates the variables `x` and `y`, which keep the robot position in the matrix.

> **Rule 38** (generate-robot-coordinates) [[**Position(**x, y**)**, bound-size]] $^{\text{generate-robot-coordinates}}$
> **proviso**. x<*bound-size* and y<*bound-size*
>
> ⇓
>
> **x : [1..n] init** [[x]] **;**
> **y : [1..n] init** [[y]] **;**

Robot's Movement Generation. The rules for the generation of transitions that represent the robot's movement are most numerous and complex. For instance, consider the semantics for the **Up()** command.

> **Rule 42** (instruction-controller) [[**Up()**, n]] $^{\text{instruction-controller}}$
>
> ⇓
>
> [] **(movement = false) & (moving = false) & (instruction =** [[n]] **) & (y - 1 != 0)**
> [[**upIsClear()**, 0, [[0]] $^{\text{get-obstacle-count}}$]] $^{\text{create-conditional-obstacle-evaluation-controller}}$
> **-> processes : (y' = y - 1) & (instruction' = instruction + 1) & (movement' = true);**

This rule outputs the elements already discussed in Sect. 4.3. The function **upIsClear()**, that is mentioned in this rule, is subsequently translated into guards that are evaluated to true if there is no obstacle in the adjacent cell that is northwards the robot current position (as described by the Rule 95 presented in the sequel).

Obstacle Evaluations' Generation. In the previous rule, we saw that some translation rules are called inside conditional expressions. This is because the condition to check is considerably complex to create by a single function. Here we have an example of a rule that complements the previous rule's condition.

Rule 95 (create-obstacle-evaluation) [[**Up()**, n, end]] $^{\text{create-obstacle-evaluation}}$

$\Downarrow$

& (((y - 1 = obstacle_ [[n + 1]] **_vertical_alignment &**
(x < obstacle_ [[n + 1]] **_left_bound & x > obstacle_** [[n + 1]] **_right_bound))**
| (y − 1 != obstacle_ [[n + 1]] **_vertical_alignment))**

[[**Up()**, n + 1, end]] $^{\text{create-obstacle-evaluation}}$

Note that the result of the previous rule is not a command by itself but part of a boolean condition as expected by Rule 42 or other.

Properties Generation. In order to illustrate how a DSL property is translated into the respective PRISM property, we consider the "`Energy`" property. The first part of the translation creates a reward. The `energy_constant` is a constant equals the average energy consumption of a robot to perform a movement. It must be set considering the particular robot model under analysis.

Rule 116 (generate-rewards) [[Props]] $^{\text{generate-rewards}}$

$\Downarrow$

rewards "energy"
[] (movement = true) : energy_constant;
endrewards

The second part of the translation outputs the associated temporal formula to be evaluated by PRISM, as follows.

Rule 120 (parse-energy) [[Props]] $^{\text{parse-energy}}$

$\Downarrow$

R"energy"=? [F instruction = total_instructions]

The previous property asks PRISM to calculate the amount of energy necessary to achieve the `total_instructions`. Note that this kind of energy consumption solution is quite simple by just associating a unit of energy to each movement of the robot. Ideally we have to employ a more realistic model. But to have a preliminary idea to compare different algorithms, this solution is enough.

4.5 Tool support

Figure 6 shows a screenshot of the integrated environment (Eclipse plugin) for authoring programs written in the proposed DSL. In this figure, one can observe six tabs: module.rob—in this tab, one can write a program in the DSL with syntax coloring and validation; module.prism—where the PRISM model automatically generated is output, every time the program is saved; module.props—here, the automatically generated Prism properties are output; robot_dsl.str—this tab specifies the DSL itself and must not to be modified; error_handling—such a tab shows error messages; and, the tab Robot DSL Map file—where the map representing the environment is input.

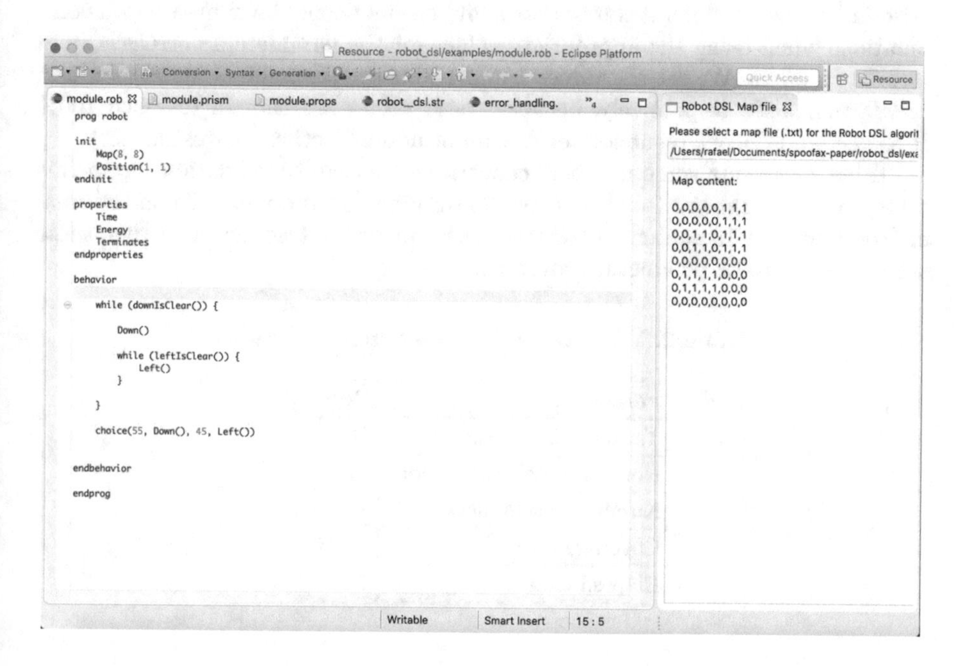

Fig. 6. DSL editor and PRISM translator

5 Validating the DSL

Ideally, translation rules must be formally proved to be sound and complete [22]. However, in the absence of a formal proof, other kinds of validations are welcome. We have performed a more pragmatic validation by checking whether correct DSL-based fragments (complete programs in our proposed DSL) create compilable PRISM specifications after applying our rules. To generate the DSL fragments, we used the sentence generator named Lua Language Generator (LGen) [10].

The LGen tool accepts as input the Extended Backus-Naur Form (EBNF) notation of the (context-free syntax of the) language of interest, and generates a Lua [14] program that produces a set of programs conforming to the EBNF. To use the LGen tool, one has to set some parameters to give a upper bound for the program generation as, for instance, limit the recursion depth of some grammar elements by defining the maximum number of recursion cycles of non-terminal elements (parameter *maxCycles*). Additionally, one has to define a coverage criteria. The available coverage criteria are.

- *Terminal Symbol Coverage*: coverage of language terminal symbols;
- *Production Coverage*: coverage of language production rules.

Recall from Sect. 4.1 that our top-level element is a *Program*. So we applied the LGen on this non-terminal. We choose *Production Coverage* as coverage criteria because it generates more tests and covers deeper level instruction combinations when using the appropriate values for the parameters *maxCycles* and *maxDerivLength*. We set the values 1 and 13 for the parameters *maxCycles* and *maxDerivLength*, respectively, because these values are enough to reach 100% coverage, and deeper recursion levels cannot add any other interesting test.

Table 2 presents some numbers concerning this sanity validation. From this table, we can see that LGen generated 33 terminal instructions, 27 non-terminal instructions and 19 sentences (or test cases) in just 7 seconds. And that these sentences get 100% production coverage.

Table 2. Values of the coverage criteria generation

Parameter	Result
Terminal instructions	33
Non-Terminal instructions	27
Number of sentences	19
Coverage rate	100%
Elapsed time	7 s

The use of LGen was very interesting because it quickly revealed some subtle issues (some tests failed!) in our translation rules . All issues have been fixed in the current version of the translation rules.

6 Evaluation

In this section we present evaluations performed on our proposed solution towards robot motion planning. We first consider the scalability of our PRISM automatically generated specifications. Thus, we consider how to use our tool to perform comparisons between different robot motion algorithms. We have used a machine with an Intel Core i7 processor (2.5 GHz), 8 GB RAM, running Windows 10.

6.1 Model's Scalability

As discussed in Sect. 4.2, the cost to analyze large models can make model checking infeasible. In the context of this work, the size of the matrix has a relevant impact in the scalability of the analysis performed by PRISM. In this section we investigate whether our proposal scales well by increasing the matrix size. We consider three different dimensions to observe the effort taken by PRISM in each case.

- A 5×5 matrix;
- A 8×8 matrix, and;
- A 10×10 matrix

These matrices correspond to rooms of dimensions $1.7\,\text{m} \times 1.7\,\text{m}$, $2.72\,\text{m} \times 2.72\,\text{m}$ and $3.4\,\text{m} \times 3.4\,\text{m}$. The calculi for the room dimensions consider each matrix cell has the shape of an iRobot vacuum (an almost $34\,\text{cm} \times 34\,\text{cm}^2$).

We used the following behavior to move the robot through the rooms in the same manner, where the initial position of the robot is (1, 1)[7] and the obstacles are placed in the upper-right section, with their left and right boundaries varying by their size in the row in which they are placed.

```
behavior
  while (downIsClear()) {
    while (rightIsClear()) {
      Right()
    }
    while (leftIsClear()) {
      Left()
    }
    Down()
  }
endbehavior
```

Table 3 presents the effort to create the Markov chains from our PRISM specifications.

Table 3. Comparison of the built models for different matrix sizes

Built model values	5×5 matrix	8×8 matrix	10×10 matrix
States	99	285	420
Transitions	99	285	420
Build time	≈2.4 s	≈5 s	≈32 s

From the previous table we can see that our translation rules try to optimize the construction of the state space by creating the minimum possible number of states and transitions. In this case we get the exact same number of states and transitions for the three matrix sizes.

Table 4 shows the results we get by analyzing the same behavior for varying matrices dimensions. Note that when the probability of completion is 100%, the number of executed instructions coincides with the energy required. Our power consumption model is linear with the number of instructions and satisfying the mission (to cover all free accessible cells of the grid), just to illustrate the concept. But we can adapt it to more realistic power consumption models.

As we can see from Table 4, the analysis effort (time in seconds) evolves exponentially with the size of the matrix. This indicates that we can only use our solution for reasonable room sizes. Otherwise the PRISM model checker can exhibit the *state explosion problem* as reported in the literature [5].

[7] The upper-left portion of the matrix.

Table 4. Comparison of the 5×5, 8×8 and 10×10 matrices

Rewards	5×5 matrix	8×8 matrix	10×10 matrix
Probability of completion	100%	100%	100%
Executed instructions	31	93	124
Time required	≈80.0 s	≈267.0 s	≈344.1 s
Energy required	31	93	124

6.2 Comparing algorithms

We compare the performance of three different traversal (cleaning) algorithms. For these algorithms, we used the following DSL fragment, which specifies a matrix of size 8×8 (a medium-sized room with $7.4\,\mathrm{m}^2$) and the same properties to be analyzed (probability to terminate, energy and time required to complete the mission).

```
prog robot
init
    Map(8, 8)
    Position(?, ?)
endinit
properties
    Terminates
    Energy
    Time
endproperties
(behavior section)
endprog
```

The initial position (`(?, ?)`) and the `(behavior section)` are specific for each algorithm and can be found in Fig. 7.

The three algorithms used the following map:

```
0,0,0,0,0,1,1,1
1,1,0,0,0,1,1,1
1,1,0,0,0,1,1,1
0,0,0,0,0,1,1,1
0,0,0,1,1,1,1,1
0,0,0,1,1,1,1,1
0,0,0,1,1,1,1,1
0,0,0,1,1,1,1,1
```

In Algorithm I, the robot initially tries to move to the right while no obstacle is found. Next, the robot attempts to follow a more complicated path. First, it tries to move one position down. Then, it tries to reach the leftmost position. If an obstacle is found, it tries to go down one cell, and tries to reach the rightmost position and the whole traversal repeats.

Algorithm II is different from the previous one because the robot starts at a different position (1, 8), and also because it uses the `choice` operator to decide the movement. The algorithm starts by going up while it can (before it reaches an obstruction) before going right by one cell, then going down while it can. This is repeated until the movement to the right is not possible. Then it goes up

```
Algorithm I (1, 1):
behavior

while (rightIsClear()) {
  Right()
}

while (downIsClear()) {

  Down()

  while (leftIsClear()) {
        Left()
  }

  Down()

  while (rightIsClear())
        {
        Right()
  }

}

endbehavior
```

```
Algorithm II (1, 8):
behavior
while (rightIsClear()) {
  While (upIsClear()) {
    Up()
  }
  Right()
  While (downIsClear()) {
    Down()
  }
}
while (upIsClear()) {
  Up()
}

choice(Left(), Right())
choice(Left(), Right())

while (rightIsClear()) {
  Right()
}
while (downIsClear()) {
  Down()
}
Left()
while (upIsClear()) {
  Up()
}

endbehavior
```

```
Algorithm III (5, 1):
behavior
while (downIsClear()) {

  while(LeftIsClear()) {
    Left()
  }

  while(RightIsClear()) {
    Right()
  }

  Down()

}

while (LeftIsClear()) {

  while (LeftIsClear()) {
    Left()
  }
  Down()

  while (RightIsClear()) {
    Right()
  }
  Down()

}
endbehavior
```

Fig. 7. Algorithms used in our experiments comparison

while it can, perform the two choices before shifting to a specific pattern until completing the cells. The path that the robot follows in the algorithm is branched in two, every time the choice is made, so, there is a 50%-50% probability of going leftward and rightward. Considering the choice is made two consecutive times, the robot has a 25% probability of completing the matrix.

Finally, Algorithm III uses a similar strategy to Algorithm I, but the robot starts in a initial position different from I (5,1). It goes left until an obstacle/boundary is reached, then goes to the right until it hits another obstacle/boundary, then goes down by one cell. After that, if a movement downwards is not possible, it moves to its left until it hits a boundary/obstacle, goes down by one cell, moves right until it hits a boundary/obstacle and down by one cell repeatedly until no movement to the left it possible, or the total number of passable cells is traversed by the robot, whatever comes first.

Table 5 gives us an idea of the effort to create the Markov chains for the three algorithms.

Table 5. Comparison of built values between models

Built model values	Algorithm I	Algorithm II	Algorithm III
States	108	261	141
Transitions	108	264	141
Building time	≈4.3 s	≈6.0 s	≈5.1 s

From Table 5 we can observe that Algorithms I and III have the same number of states and transitions, whereas Algorithm II has more states and transitions.

This is because the Algorithms I and III do not have probabilistic choices, thus generating a straightforward model with an optimal path by design.

In Table 6 we can see that only Algorithm II cannot complete its mission with a 100% guarantee. And this naturally implies an infinity amount of energy to accomplish this task. The other two algorithms need the same amount of energy as the number of instructions because we have associated one unit of energy to each executed instruction.

Table 6. Comparison of properties between algorithms

Reward	Algorithm I	Algorithm II	Algorithm III
Probability of completion	100%	25%	100%
Executed instructions	34	43	45
Time required	≈96.0	≈140.0	≈148.0
Energy required	34	∞	45

6.3 A More Realistic Environment

In the previous section our goal was to compare algorithms on simple maps. In this section our goal is to compare three navigational algorithms on a more realistic environment. The following map (left-hand side) captures abstractly an environment where a bed is placed in the upper right corner, an armchair is placed somewhere in the middle of the room. Moreover, a rack is placed in the lower part of the map preserving a certain free area between the rack and the down corners. In the right-hand side we display the corresponding DSL part for this map and the properties we are interested to analyze. The initial position is shown together with the respective behaviors in Fig. 8.

```
0,0,0,0,0,1,1,1
0,0,0,0,0,1,1,1
0,0,1,1,0,1,1,1
0,0,1,1,0,1,1,1
0,0,0,0,0,0,0,0
0,1,1,1,1,0,0,0
0,1,1,1,1,0,0,0
0,0,0,0,0,0,0,0
```

```
init
    Map(8, 8)
    Position(?, ?)
endinit

properties
    Time
    Energy
    Terminates
endproperties
```

The behavior part of the DSL programs are

From Fig. 8, Algorithm I starts at position (1, 1). It tries to go to the right-hand side of the grid as much as possible. When it is not possible, it tries to gown down one cell initially and then it tries a left-down-right-down movement as much as possible. When it is no longer possible to do this movement, it performs a 50-50% probabilistic choice between going up and left. After this choice, it tries to perform to movement left as much as possible, up as much as possible, right as much as possible, and down as much as possible.

Algorithm II is quite similar to Algorithm I, although its starting position is (1, 8). The difference is mainly related to its starting position. From it, this

Algorithm I (1, 1):	Algorithm II (1, 8):	Algorithm III (8, 5):

<pre>behavior
 	
  while (rightIsClear())
        {
    Right()
  }
  if (downIsClear()) {
    Down()
  }
  while (downIsClear()) {
    while (leftIsClear())
          {
      Left()
    }
    if (downIsClear()) {
      Down()
    }
    while (rightIsClear()
) {
      Right()
    }
    if (downIsClear()) {
      Down()
    }
  }
 	
  choice(Up(), Left())
 	
  while (leftIsClear()) {
    Left()
  }
  while (upIsClear()) {
    Up()
  }
  while (rightIsClear())
        {
    Right()
  }
  while (downIsClear()) {
    Down()
  }
endbehavior</pre>
	<pre>behavior
	  while (upIsClear()) {
	    while (rightIsClear()) {
	      Right()
	    }
	    if (upIsClear()) {
	      Up()
	    }
	    if (upIsClear()) {
	      while (leftIsClear())
	            {
	        Left()
	      }
	    }
	    if (upIsClear()) {
	      Up()
	    }
	  }
	  while (leftIsClear()) {
	    Left()
	  }
	  while (downIsClear()) {
	    Down()
	  }
	  while (upIsClear()) {
	    Up()
	  }
	  while (rightIsClear()) {
	    Right()
	  }
	  while (downIsClear()) {
	    Down()
	  }
	  while (leftIsClear()) {
	    Left()
	  }
	  choice(Up(), Right())
	  while (upIsClear()) {
	    Up()
	  }
	  choice(Right(), Down())
	  while (rightIsClear()) {
	    Right()
	  }
	endbehavior</pre>
		<pre>behavior
		  while (downIsClear()) {
		    Down()
		  }
		  choice(Left(), Right())
		  while (upIsClear()) {
		    Up()
		  }
		  choice(Left(), Right())
		  while (downIsClear()) {
		    Down()
		  }
		  while (leftIsClear()) {
		    Left()
		  }
		  while (upIsClear()) {
		    Up()
		  }
		  while (rightIsClear()) {
		    Right()
		  }
		  while (downIsClear()) {
		    Down()
		  }
		  while (upIsClear()) {
		    Up()
		  }
		  choice(Left(), Down())
		  while (leftIsClear()) {
		    Left()
		  }
		  choice(Right(), Down())
		  while (downIsClear()) {
		    Down()
		  }
		  while (rightIsClear()) {
		    Right()
		  }
		endbehavior</pre>

Fig. 8. Algorithms used in our experiments comparison

algorithm tries to move the robot upward in a right as much as possible, one cell up, left as much as possible, and one cell up. At the point where it can achieve, it tries to go left as much as possible, then down as much as possible and then returning to its starting point (going up), then it turns to the right as much as possible. And then it tries to follow to go down and left, independently, as much as possible. Similarly to Algorithm I, it chooses between two possible directions, up and right, fairly (50–50%). Thus it tries to go up, and makes another fair choice between right and down. At this point, it tries to follow rightward until it finishes its movement.

Finally, we have Algorithm III, which although it has the smallest code, it is the most random from the three algorithms. It has four choices. Starting from position (8, 5), it tries a downward movement, followed by a left or right fair choice. Then it tries an upward movement, followed again by a left or right fair choice. Now, from this position, this algorithm tries as much as possible and independently to perform the movement down-left-up-right-down-up. When it reaches some obstacle, it makes another fair choice between left or down, and it tries to go to the left as much as possible. When it can no longer perform the previous movement, it performs the final fair choice between right and down, and tries to follow downward and then to the right.

Table 7 gives us an idea of the effort to create the Markov chains for the three algorithms.

Table 7. Comparison of built values between models

Built model values	Algorithm I	Algorithm II	Algorithm III
States	258	290	1490
Transitions	259	293	1505
Building time	$\approx$15 s	$\approx$18 s	$\approx$22 s

From Table 7 we can observe that Algorithms I and II have similar number of states and transitions, whereas Algorithm III has almost five times more states and transitions. This is because the Algorithms I and II have few probabilistic choices (one and two, respectively) than Algorithm III, which has 4 probabilistic choices. The places where there choices are located also influence the randomness of the algorithm. Fortunately, the time to build Algorithm III does not even double.

In Table 8 we can see the effort each algorithm takes to calculate the expected properties. Note that the high degree of randomness in Algorithm III prejudices its performance to complete the mission. This is the main reason Neato has sensors to "see the environment (grid)" before to calculate its path: this turns the navigational algorithm more deterministic. All three algorithms cannot assure mission completeness and in view of this, all of them would required infinite energy to finish their job. On the other hand, they execute a number of finite instructions because after these amounts, these algorithms cannot execute more instructions.

Table 8. Comparison of properties between algorithms

Reward	Algorithm I	Algorithm II	Algorithm III
Probability of completion	50%	25%	6.25%
Executed instructions	44	51	52
Time required	$\approx$150	$\approx$175	$\approx$180
Energy required	∞	∞	∞

7 Related Work

We could not find in the literature similar works that perform stochastic verification of floor cleaning robots using a DSL as the input notation as our work proposes. Unlike our work, several works in the literature perform experiments for evaluating the robots. In general, the experiments consists of running the robot, recording its behavior and then analyzing the recordings to assess the robot performance [20,21,27]. For instance, the work [20] introduces a method and dedicated hardware apparatus for filming and analyzing cleaning robots. The work [27] also develops its own methods and tools to evaluate a programmable Roomba robot.

There exist works that use probabilistic model checking for the verification of robot swarms [4,17]. In [17] the authors use the PRISM model checker to verify forage robot swarms that collaborate on the search of food and another kind of resources. In this work, there is no model for an individual robot: the swarm is modeled as a whole using a Discrete-Time Markov Chain (DTMC) that represents the cycle of a forage robots swarm (resting, searching, grabbing, depositing food and returning home). In another work [4], the authors also use PRISM to verify robot swarms by following a property-driven design approach. Similar to our work, [4,17] use PRISM to model an abstract (and platform independent model) of the robot and its map, and Probabilistic Computation Tree Logic (PCTL) formulas to specify the properties of interest to be verified in PRISM. An important difference is that our approach uses a DSL for specifying the robot behavior, while in [4,17] robot behavior is specified directly in PRISM notation.

There are several approaches in the literature that use regular model checkers (that do not consider probabilities) for the verification of robots [23,28] in other domains than floor cleaning robots. For example, the work [23] develops an extension for the Java PathFinder tool to perform an exhaustive behavioral check of the source code of a line follower robot considering the interactions of the robot with the physical environment. The work [28] also proposes an approach that uses a model checker for verifying the behavior of a set of robots that collaborate in an industrial setting. Different from our approach, that inputs a model written in a high-level DSL, the approach in [28] consists on abstracting the robots controller code into a formal model that is the input for the Spin model checker [12].

An alternative approach for the verification of robots is to synthesize the robot control program. Robot motion (or path) planning aims at automatically finding the robot movements required for performing a well-defined task. Path planning has been used in the domain of cleaning robots [11,13,29], as well as the planning of robots that perform more general tasks [9,19,25]. The work [11] is one of the pioneer works on the synthesis of control programs for cleaning robots; it proposes a semi-automatic approach for the generation of a robot motion program that ensures the coverage of an area to be cleaned considering two specific robot models. A more recent work is [29] that proposes an approach for solving the Path Planning Coverage Region using a genetic algorithm based on an evolutionary approach. The purpose of our work is to verify robot algorithms, not to synthesize them. Additionally, we are interested in verifying programs written in a DSL whose notation is similar to an imperative program language. The works [11,19,25,29] output the robot algorithm in mathematical notations that can be less readable than the proposed DSL. For instance, the work [25] needs a Markov mathematical presentation to exhibit its results, and the work [19] expresses the motion plan as an automaton.

8 Conclusion

This work presents a practical tool to programmers of cleaning robots: a way of estimating the necessary time and energy to complete a mission. Usually a cleaning robot is just set to run for some time but there is no clear idea about its success of cleaning the intended environment beforehand nor the amount of energy needed. This is one of the reasons industry has created the auto-charger feature and of using traditional motion planning algorithms [24].

Our proposal is feasible due to the advances in probabilistic model checking. We can overcome simulation-based analysis and bring more certainty about the robot behavior on different environments. By investigating algorithms with such a confidence we also provide a more ecological solution because we can easily look for designs that require less power consumption.

We did not follow a completely formal development of our tool but we performed a preliminary investigation about its trustworthiness using a grammar generation to check whether our DSL to PRISM translator provided PRISM Model Checker specifications that could be loaded and analyzed without raising issues. Obviously we could not check whether the proposed PRISM specifications make sense in the formal semantic perspective.

We have presented the results of the performance of three simple algorithms using our approach. Although the analysed algorithms are simple, the results presented so far illustrate how our approach can be useful to compare different algorithms. The unique effort of the robot developer is to describe the algorithm in the DSL as well as stating the properties to be analysed. To give an idea of our proposed DSL and analysis, we have also presented a more realistic study where the environment abstractly captured some furniture in some room. Due to the positions of the obstacles and the algorithms were not so intelligent, all of them cannot complete their mission, taken infinite energy to do so.

To improve the quality of the proposed approach to the motion planning algorithm evaluation, there are some points that can be given a more polished design and implementation.

A first future work is to compare our abstract analysis with the behavior of a real robot; for instance, the iRobot Create equipment [6]. Another future direction is to consider the translation from continuous motion planning models to discrete PRISM Model Checker (PRISM) specifications. Industry uses continuous models that are based on the models of robot sensors. We intend to avoid the use of traditional sensors as much as possible with the use of a camera and image processing.

In this work, we use navigational matrices of hypothetical scenarios. A future work may consider matrices of real environments. To accomplish this, we intend to use image processing algorithms applied to pictures of such environments.

Related to the PRISM model, we may apply the *renaming* directives of the language to reduce the underlying model's size significantly, generating a less populated target file that is better parsed by the PRISM support tool. Additionally, the *formula* element can be used to specify the evaluation of the instructions so they can be reused in the guards instead of expanding them as the translation rules do.

Another future work is to improve the DSL's base syntax to accept some conveniences, such as multiple evaluations in a single control structure, support for user-created variables, support for n factor choices, enabling a chain of choices to be accepted within other choices, and the optimization of translation rules and its calls during the target model generation.

A final future work would be a completely formal approach to investigate our translator and the semantic means of the resulting analysis. To do that we will need to work directly with Markov chains and processes, the semantics behind the PRISM model checker.

Acknowledgements. This work was funded by CNPq, grants 442859/2014-7 and 302170/2016-2. This work was partially supported by the National Institute of Science and Technology for Software Engineering (INES - http://www.ines.org.br), funded by CNPq and FACEPE, grants 573964/2008-4 and APQ-1037-1.03/08.

References

1. Angle, C., Greiner, H., Brooks, R.: iRobot: the robot company (1991). http://www.irobot.com
2. de Araújo, R.P.: Probabilistic analysis applied to robots. Master's thesis, Centro de Informática (UFPE), Brazil (2016)
3. Baier, C., Katoen, J.P.: Principles of Model Checking (Representation and Mind Series). The MIT Press, Cambridge (2008)
4. Brambilla, M., Brutschy, A., Dorigo, M., Birattari, M.: Property-driven design for robot swarms: a design method based on prescriptive modeling and model checking. ACM Trans. Auton. Adapt. Syst. **9**(4), 17:1–17:28 (2014). http://doi.acm.org/10.1145/2700318
5. Clarke Jr., E.M., Grumberg, O., Peled, D.A.: Model Checking. MIT Press, Cambridge (1999)
6. iRobot Corporation: iRobot create 2 programmable robot. http://www.irobot.com/About-iRobot/STEM/Create-2.aspx
7. Dehnert, C., Junges, S., Katoen, J.P., Volk, M.: A storm is coming: a modern probabilistic model checker. arXiv preprint arXiv:1702.04311 (2017)
8. Filieri, A., Ghezzi, C., Tamburrelli, G.: Run-time efficient probabilistic model checking. In: Proceedings of the 33rd International Conference on Software Engineering, ICSE 2011, pp. 341–350. ACM, New York (2011). http://doi.acm.org/10.1145/1985793.1985840
9. Galceran, E., Carreras, M.: A survey on coverage path planning for robotics. Robot. Auton. Syst. **61**(12), 1258–1276 (2013). http://www.sciencedirect.com/science/article/pii/S092188901300167X
10. Hentz, C.: Automatic generation of tests from language descriptions. Master's thesis, Federal University of Rio Grande do Norte, Natal, Brazil (2010)
11. Hofner, C., Schmidt, G.: Path planning and guidance techniques for an autonomous mobile cleaning robot. Robot. Auton. Syst. **14**(2), 199–212 (1995). http://www.sciencedirect.com/science/article/pii/092188909400034Y. Research on Autonomous Mobile Robots

12. Holzmann, G.: Spin Model Checker, the: Primer and Reference Manual, 1st edn. Addison-Wesley Professional (2003)
13. Hu, G., Hu, Z., Wang, H.: Complete coverage path planning for road cleaning robot. In: 2010 International Conference on Networking, Sensing and Control (ICNSC), pp. 643–648 (2010). https://doi.org/10.1109/ICNSC.2010.5461585
14. Ierusalimschy, R., de Figueiredo, L.H., Filho, W.C.: Lua-an extensible extension language. Software: Practice and Experience **26**(6), 635–652 (1996). https://doi.org/10.1002/(SICI)1097-024X(199606)26:6⟨635::AID-SPE26⟩3.0.CO;2-P
15. Katoen, J.P., Zapreev, I.S., Hahn, E.M., Hermanns, H., Jansen, D.N.: The ins and outs of the probabilistic model checker MRMC. Perform. Eval. **68**(2), 90–104 (2011). https://doi.org/10.1016/j.peva.2010.04.001
16. Kochenderfer, M.J., Amato, C., Chowdhary, G., How, J.P., Reynolds, H.J.D., Thornton, J.R., Torres-Carrasquillo, P.A., Üre, N.K., Vian, J.: Decision Making Under Uncertainty: Theory and Application, 1st edn. The MIT Press (2015)
17. Konur, S., Dixon, C., Fisher, M.: Analysing robot swarm behaviour via probabilistic model checking. Robot. Auton. Syst. **60**(2), 199–213 (2012). http://www.sciencedirect.com/science/article/pii/S0921889011001916
18. Kwiatkowska, M., Norman, G., Parker, D.: PRISM 4.0: verification of probabilistic real-time systems. In: Gopalakrishnan, G., Qadeer, S. (eds.) Proceedings of 23rd International Conference on Computer Aided Verification (CAV 2011). LNCS, vol. 6806, pp. 585–591. Springer (2011)
19. Lahijanian, M., Almagor, S., Fried, D., Kavraki, L.E., Vardi, M.Y.: This time the robot settles for a cost: a quantitative approach to temporal logic planning with partial satisfaction. In: AAAI, pp. 3664–3671 (2015)
20. Palacin, J., Palleja, T., Valganon, I., Pernia, R., Roca, J.: Measuring coverage performances of a floor cleaning mobile robot using a vision system. In: Proceedings of the 2005 IEEE International Conference on Robotics and Automation, pp. 4236–4241 (2005). https://doi.org/10.1109/ROBOT.2005.1570771
21. Palleja, T., Tresanchez, M., Teixido, M., Palacin, J.: Modeling floor-cleaning coverage performances of some domestic mobile robots in a reduced scenario. Robot. Auton. Syst. **58**(1), 37–45 (2010). http://www.sciencedirect.com/science/article/pii/S0921889009001171
22. Sampaio, A.: An Algebraic Approach to Compiler Design, vol. 4. World Scientific (1997)
23. Scherer, S., Lerda, F., Clarke, E.M.: Model checking of robotic control systems. In: Proceedings of ISAIRAS 2005 Conference, pp. 5–8 (2005)
24. Thrun, S., Burgard, W., Fox, D.: Probabilistic Robotics (Intelligent Robotics and Autonomous Agents). The MIT Press (2005)
25. Toit, N.E.D., Burdick, J.W.: Robot motion planning in dynamic, uncertain environments. IEEE Trans. Robot. **28**(1), 101–115 (2012). https://doi.org/10.1109/TRO.2011.2166435
26. Tribastone, M.: The PEPA plug-in project (2009). https://doi.org/10.1109/QEST.2007.34
27. Tribelhorn, B., Dodds, Z.: Evaluating the roomba: a low-cost, ubiquitous platform for robotics research and education. In: ICRA, pp. 1393–1399. IEEE (2007). http://dblp.uni-trier.de/db/conf/icra/icra2007.html#TribelhornD07
28. Weißmann, M., Bedenk, S., Buckl, C., Knoll, A.: Model checking industrial robot systems, pp. 161–176. Springer, Heidelberg (2011). https://doi.org/10.1007/978-3-642-22306-8_11

29. Yakoubi, M.A., Laskri, M.T.: The path planning of cleaner robot for coverage region using genetic algorithms. J. Innovation Digital Ecosyst. **3**(1), 37–43 (2016). http://www.sciencedirect.com/science/article/pii/S2352664516300050
30. Younes, H., Kwiatkowska, M., Norman, G., Parker, D.: Numerical vs. statistical probabilistic model checking. Int. J. Software Tools Technol. Transf. (STTT) **8**(3), 216–228 (2006)
31. Younes, H.L.S., Kwiatkowska, M., Norman, G., Parker, D.: Numerical vs. statistical probabilistic model checking: an empirical study, pp. 46–60. Springer, Heidelberg (2004). https://doi.org/10.1007/978-3-540-24730-2_4

From Petri Nets to UML: A New Approach for Model Analysis

Lila Meziani[1(✉)], Thouraya Bouabana-Tebibel[1], Lydia Bouzar-Benlabiod[1], and Stuart H. Rubin[2]

[1] Laboratoire de la communication dans les systèmes informatiques, Ecole nationale Supérieure d'Informatique, BP 68M, 16309 Oued-Smar, Algiers, Algeria
{l_meziani,t_tebibel,l_bouzar}@esi.dz

[2] Space and Naval Warfare Systems Center Pacific, San Diego, CA 92152-5001, USA
stuart.rubin@navy.mil

Abstract. UML is a semi-formal notation largely adopted in the industry as the standard language for software design and analysis. Its imprecise semantics prevents any verification task. However, a formal semantics can be given to UML diagrams, for instance, through their transformation to models with a formal semantics, such as Colored Petri Nets (CPN).

Colored Petri nets are an efficient language for UML state machine formalization and analysis. In order to assist the UML modeler in understanding the report generated by the Petri net tool, we propose a method to construct UML diagrams from the returned report. A case study is given to illustrate the proposed approach.

Keywords: Model checking · Petri nets · State machine · UML

1 Introduction

UML is a semi-formal notation largely adopted in the industry as the standard language for software design and analysis. Its imprecise semantics prevents any verification task. However, a formal semantics can be given to UML diagrams, for instance, through their transformation to models with a formal semantics, such as Colored Petri Nets (CPN).

The UML community is divided into two categories: those concerned with UML modeling and those interested in UML formalization. It follows that the first community often does not understand the results returned by the formal analysis and validation tools. Our approach gives a solution to this problem by producing UML diagrams from the analyzed CPN models. These diagrams provide the analysis results in a form legible for the UML designers.

Our proposed solution takes as input a CPN model derived from a UML state machine (SM). We perform the CPN analysis using the model checker CPNTools

T. Bouabana-Tebibel et al. (Eds.): IEEE IRI 2017, AISC 838, pp. 52–68, 2019.
https://doi.org/10.1007/978-3-319-98056-0_3

[16]. The results of the analysis are then interpreted and returned to the UML modeler in a graphical form, using sequence and object diagrams.

The rest of the paper is organized as follows: First, we present some previous works which propose formal semantics to some UML diagrams. In the third section, we give a background on the UML and Petri Net modeling as well as SM formalization. We detail our contribution in the fourth and five section. The last section concludes the paper.

2 Previous Work

Several approaches have already studied the semantics of the different UML models. In this section we enumerate some of the principal papers that deal with model formalization. We focus on the approaches trying to give a formal semantics to some UML diagrams.

The approach detailed in [8,9] takes into account five basic diagram types, namely class, object, state, sequence, and collaboration diagrams. The authors use graph transformation method to obtain the formal model. The graph transformation system consists of graph transformation rules and a so-called system state graph. This graph represents the state of the modeled system at a given point in time. The changes of the system state during an execution of the model are simulated by the application of graph transformation rules on the system state graph.

In [6], the authors propose an algorithm that formally transforms dynamic behaviors of systems expressed in UML into their equivalent Colored Petri Nets models using the graph transformation algorithm. They focus on the UML statechart and collaboration diagrams. The Meta-Modeling tool AToM3 is used. Their objective is to detect behavioral inconsistencies. INA Petri Net analyzer is used to verify the obtained CPN. In INA Petri Net analyzer, the CPN model is defined by a textual description and the analyst must be an expert of this tool. In this work, the analysis phase has not been detailed.

Hölscher et al. [4] provides a framework for an automatic translation of a UML model (class, object, statechart, sequence, collaboration diagrams) into a graph transformation system.

The approach detailed in [7] offers a graphical formal approach to specifying the behavioral semantics of statechart diagram using graph transformation techniques. It supports many advanced features of statechart, such as composite states, firing priority, history, junction, and choice. The authors automatically derive a graph grammar from a state machine. The execution of a set of non-conflict state transitions is interpreted by a sequence of graph transformations.

In [13] authors propose an algorithm that translates the statechart diagram to flat state machines. These state machines are then converted to an Object Net Model which is a generic form of an object Petri net [10]. Then the UML collaboration diagrams are used to connect these object models to derive a single CPN for the modeled system. For the analysis of the CPN model, CPNTools is used. But, no information is given about the result analysis.

The method used in [3] is an algorithm that translates the statechart diagram into a HCPN (Hierarchical Colored Petri Net). They used CPN-AMI environment for the Petri net modeling and simulation. Model checking is also achieved using CPN-AMI [11]. The PROD tool [14] is used to check some properties expressed by the Linear-Time Temporal Logic. In this work, the formal model is analyzed by PROD, and the analysis results are expressed in a textual form. The analyst must understand the Petri nets language as presented by PROD.

In [1] the authors implement a verification tool that translates UML diagrams into the input language of Zot [12] which is a bounded model satisfiability checker, thus providing users with a complete specification and verification environment. As to the authors of [15], they build a platform for the automatic transformation of a statechart to colored Petri nets.

All these works deal with the transformation or the rewriting of some UML diagrams to a formal model in order to check the correctness of the modeled system. The analysis is done in the target domain and the designer must learn the formal language to understand the analysis results.

A parallel work to our approach is given by [17]. In this work, the authors present an approach that translates CPN models obtained from business processing modeling notation to UML models. The mapping process let the generation of a use case and an activity diagrams.

3 Background and Case Study

In order to explain the core mechanism of our approach, we will briefly introduce the formal definition of CPN and some SM concepts. Next, we recall the main transformation results of SM into Object Petri Nets previously defined in [2].

3.1 Colored Petri Nets

Petri nets are a mathematical tool used for describing the system dynamics. Colored Petri nets (CPNs) [5] are an extension of Petri Nets allowing the use of data type and modularity. Therefore, CPN are more convenient for creating structured and clearer models. Colored Petri nets are used in a wide range of industrial application.

Formally, a colored Petri Net can be defined by a tuple $CPN = (\mathrm{P}, \mathrm{T}, \mathrm{A}, \sum, \mathrm{V}, \mathrm{C}, \mathrm{G}, \mathrm{E}, M_0)$ such that:

- P is a finite set of *places*,
- T is a finite set of *transitions* such that $P \cap T = \emptyset$,
- A is finite set of *arcs* such that: $P \cap T = P \cap A = T \cap A = \emptyset$,
- $\sum$ is a finite set of non-empty types, called color sets,
- V is a finite set of typed variables such that $\forall\, v \in T$, Type[v] $\in \sum$,
- C is a color function that assigns a color set to each place. It is defined from $P \rightarrow \sum$,
- G is a guard function that assigns a guard to each transition t. It is defined from $T \rightarrow$ *expressions* such that: Type (G(t)) = Bool, $\forall\, t \in T$,

- E is an arc expression function that assigns an arc expression to each arc a. It is defined from $A \rightarrow$ *expressions* such that: Type (E(a)) = C$(p)_{MS}$; p is the place connected to the arc a and MS denotes the *multiset*,
- M_0 is an initialization function that assigns an initialization expression to each place p. It is defined from $P \rightarrow$ *expressions* such that: Type[$(M_0(p))$]= C$(p)_{MS}$.

Notes:

- *expressions* is used to represent the set of expressions provided by the net inscription language. Net inscriptions are arc expressions, guards, colors sets and initial marking.
- Arcs expressions specify tokens that are added or removed by transitions.

3.2 Semantics of UML SMs

A state machine models a class behavior. It is composed of states linked by transitions. Each state can have entry/exit/do activities which are executed when entering/exiting/being in a state, respectively. An object changes its state when receiving or generating events. The generated events appear either on transitions or at the input or exit of states. They are noted on Table 1 by *evt*. The received events are noted *trigger* and appear on transitions (row 3, Table 1).

The state machine is formalized as done in [2]. The main results of the transformation approach are summarized in Table 1.

Table 1. Transformation of SM constructs into OPN

#	SM	PN	#	SM	PN
1	s	p	5	exit /evt	Link
2	t		6	entry/evt	
3	trg	Link	7	/ evt	
4	do/act	act doAct	8		final

Each state machine transformed into an Object Petri Nets (OPN)[10]. All derived Petri Nets are connected by the *Link* place. Each state is mapped to a Petri net place; each SM transition is mapped to a Petri net transition related to an input arc and output arc. For a generated event, an arc from a generic place, noted *Scenario*, to the transition where it occurs is added. A second arc from this transition to the place *Link* is also added. For an event of type trigger, an arc from the place*Link* to the triggered transition is added, and the do action is mapped to a transition-arc-place (Table 1).

3.3 Case Study

To illustrate the transformation mechanisms and those proposed in this paper, we take a case study on a peer to peer system.

A peer to peer system allows for sharing applications or resources between workstations. Each workstation acts as both a client and server.

A server can have the states *Idle*, *okconnection*, *okdisconnection* or *downloading.* When the server has no task, its state is *Idle.* This state becomes *okconnection* or *okdisconnection* when it receives a request for connection or disconnection. The state *downloading* is reached when a request for information is launched by a peer. On the other hand, a client can have the states, *connection, connected, deconnected, transmission* or *reception.* Thus, a client, for connection, sends firstly a request for the server. As soon as the request is accepted by a reception of *(okconnection).* Its state changes to *connected.* In this state, it can receive information from the server and in this case its state becomes *reception*, the received information is treated. It goes to *transmission* state, when sending information as a response to a request for information. It can also go to *deconnection* state after receiving the confirmation of deconnection *(okdeconnection)* following its request for deconnection. Figure 1 refers to the SM diagram of a server, and the corresponding Petri nets model is shown in Fig. 3. The SM diagram of a client is shown by Fig. 2 and its CPN shown by Fig. 4.

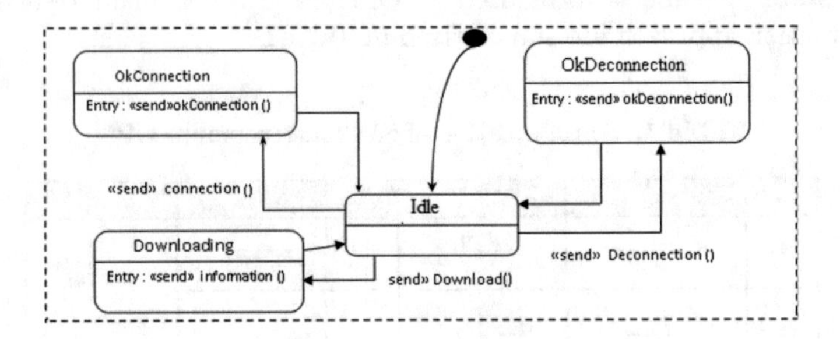

Fig. 1. State machine of a server

3.4 From OPN to HCPN

The application of the derivation rules produce an OPN, which is an abstract net model in the sense that only structured connections of net elements are considered. So, after generating the OPN system, the second step consists to enrich this abstract model with detail related to tool specific notation, such as defining color types, arc inscriptions, guards for net transitions.

In this work, we have chosen CPNTools as the underlying engine to support analysis and simulation of UML diagrams.

In CPNTools the input model is expressed in hierarchical CPN (HCPN, which extend CPN with notion of subpages and superpages). In HCPN, the model is

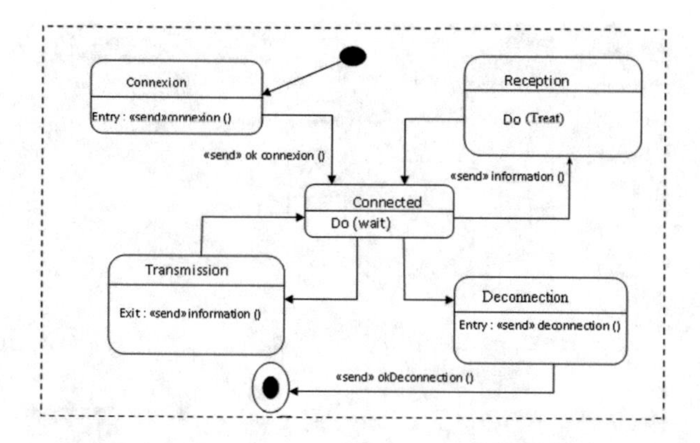

Fig. 2. State machine of a peer

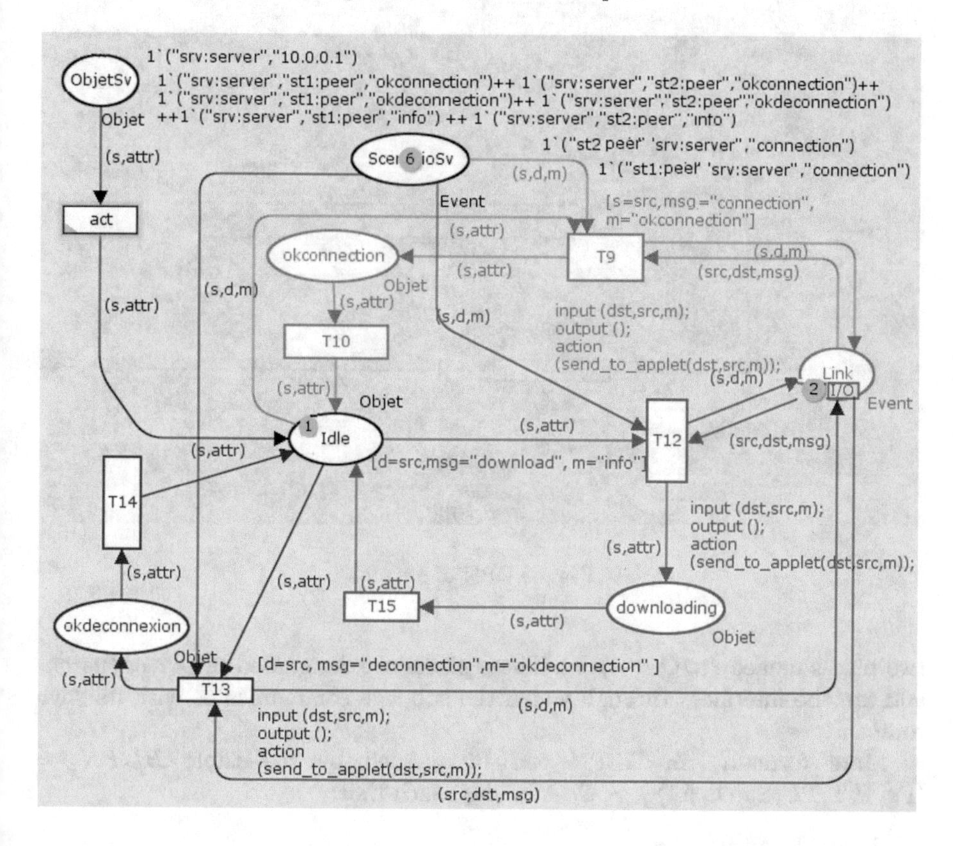

Fig. 3. CPN of a server

presented in pages. A page that contains substitutions transitions is called super-page. When a CPN-net uses a substitution transition, the subnet it represents is kept in a page named subpage (Fig. 5). Each substitution transition is related to

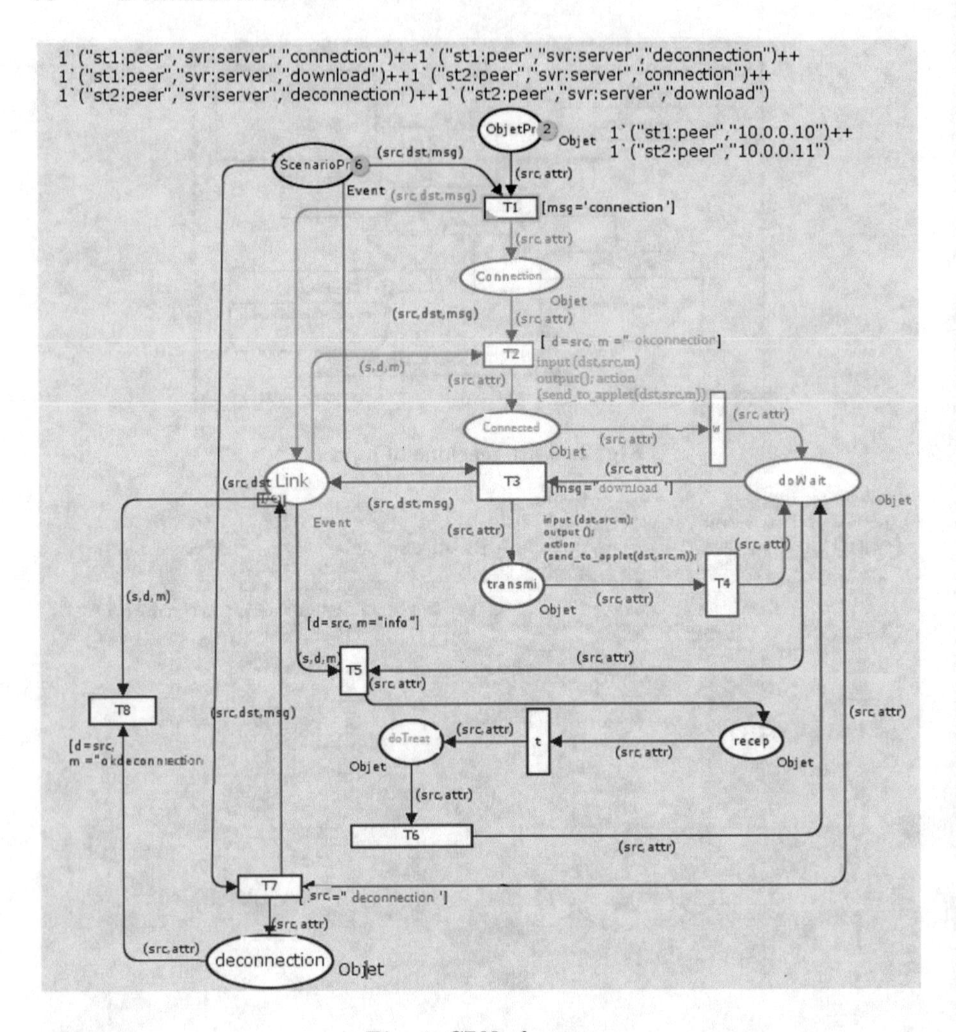

Fig. 4. CPN of a peer

two places named $In/Out - port$. These places are designated by $placesfusions$ and are the interfaces through which the subpage communicates with its environment.

More formally, an HCPN can be defined by the tuple $HCPN = (\mathrm{Pg}, \mathrm{P}, \mathrm{T}, \mathrm{PF}, \mathrm{SubT}, \mathrm{A}, \sum, \mathrm{V}, \mathrm{C}, \mathrm{G}, \mathrm{E}, M_0)$ such that:

- Pg is a finite set of *pages*,
- PF is a finite set of *place fusions*,
- $SubT$ is a finite set of *substitution transitions*,
- P, T, V, A, C, $\sum$, G, E and M_0 have the same definition as in CPN.

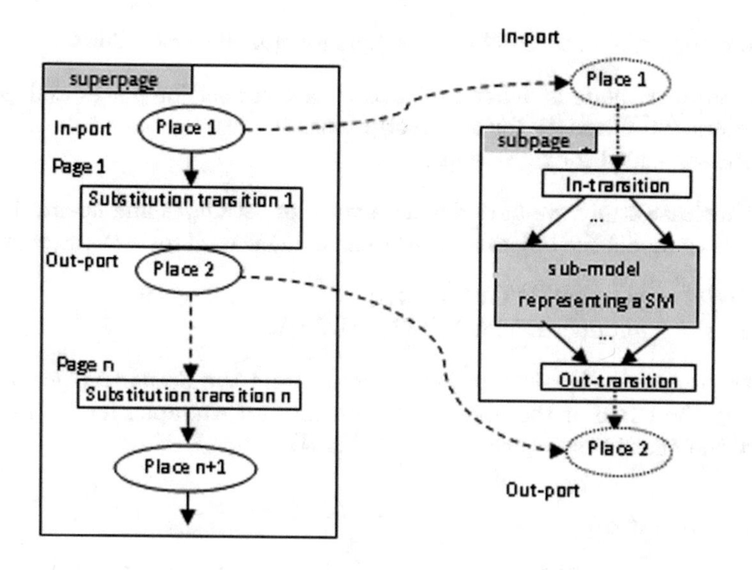

Fig. 5. Structure of the HCPN

Note:

- Fusion places form the interface through which a page communicates with its environment.

We created a superpage in which we have for each OPN a substitution transition that is linked to a subpage representing the OPN model.Then, each substitution transition is linked to the place *Link* that we used for modeling the transmission of messages. This place is a fusion-type place, it receives or sends tokens from related subpages (Fig. 6).

For the peer-to-peer model, the defined superpage, has two substitution transitions *Server* and *Peer* referring to subpages *Server* and *peer* respectively as can be seen from the tag on the bottom of a substitution transition (Fig. 6).

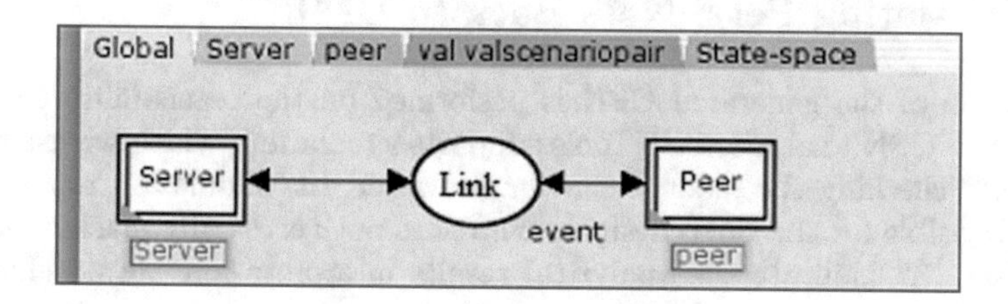

Fig. 6. Super page of peer-to-peer system

The mapping of each OPN to subpage model is done as follows:

1. A declaration node is defined for declaring color set for places and defining variables and functions for arc inscriptions.
2. Guard are added for transitions.

For our case study, we have defined two color set *Objet* and *event*. The first one serves to model the object and the second one is used for a message or event:

colset *Objet* = product DATA * DATA;
colset *Event* = product DATA * DATA *DATA.

For each transition that has an incoming arc from the *Scenario* place, a guard specifying the name of the event is defined. So, for example, for *t*1 transition, we have the guard *msg* = "*connection*" (Fig. 4).

3.5 Initialization

The execution of a CPN model requires its initialization. The places to be initialized are the *Object* and *Scenario* places of each model as well as the *Link* place. Two token types are defined, *Objet* and *Event*. The Objet tokens model the system objects. They are structured in two fields: the name of the object (obj:Class) and its attributes (atr). The Event tokens model the system events. They are composed of 3 fields: the sender (src), receiver (dst) and sent message (msg). As shown in Fig. 3, the model server contains one token in the place *Idle*: 1'("srv:server","10.0.0.1") and six tokens in the place *Scenario*:

1'("srv:server","st1:peer","okconnection")++
1'("srv:server","st2:peer","okconnection")++
1'("srv:server","st1:peer", "info")++
1'("srv:server","st2:peer","info")++
1'("srv:server","st1:peer", "okdeconnection")++
1'("srv:server","st2:peer","okdeconnection").

4 Translating Petri Nets Back to UML

Verification of the generated CPN is performed on the reachability graph. In addition to CPN analysis, CPNTools supports simulation, which we use to help a designer checking the model behavior. However, the simulation report is not comprehensible for the UML designer who does not necessarily master the CPN formalism. To facilitate the analytical results interpretation, we developed an application that: (1) returns the execution trace using a sequence diagram, (2) shows the model state, at any time of the simulation, through an object diagram.

4.1 Sequence Diagram Construction

A sequence diagram shows the exchange of messages between lifelines. Each lifeline specifies an object. In CPN, the messages are modeled by tokens of *Event* type. To produce a sequence diagram from the execution of the CPN Model, we attach a code segment to the CPN transitions. A code segment consists of a piece of sequential CPN ML code that is executed when the corresponding transition occurs in the simulation of the CPN model. As an example, the transition 'T9' (Fig. 3) has the following code segment attached:

```
input(dst,src,m);
output();
action
(send-to-applet(dest,src,m))
```

The code segment is attached only to transitions that have an incoming arc from the *Scenario* place or those which are linked to the *Link* place. It uses the function *send-to-applet* that we defined to send data to our application. A data packet is received from CPNTools at each execution of the code segment. The sender *(src)*, receiver *(dst)* and message *(msg)* fields are extracted from the received data. Next, an event is created, on the sequence diagram, from the source column to the target column labeled with the values bound to *src* and *dst.* Algorithm 1 outlines these steps.

Algorithm 1. Algorithm produce SequenceDiagram (CPN)

Begin

- for each received message *mssg* in the form (src, dest, msg)
  - Add a column C_i labeled with the bounded value to the variable src
  - Add a column C_{i+1}labeled with the bounded value to the variable dst
  - create an arc from C_i to C_{i+1} labeled with the bounded value to msg

End

The application of the Algorithm 1 to the example shown in Fig. 3 produces the sequence diagram illustrated by Fig. 7. This Figure shows three objects (st1: Peer, st2:Peer and srv:Server) interacting with each other. The horizontal arrow means that the peer object generates a new message(event) called connection() and sends this new event to the server Object. The solid rectangle means that the event *okConnection*() triggers a transition of the Peer object.

4.2 Object Diagram Construction

An object diagram describes a system state at a given moment in time. It consists of objects described with name and attribute values as well as links between the

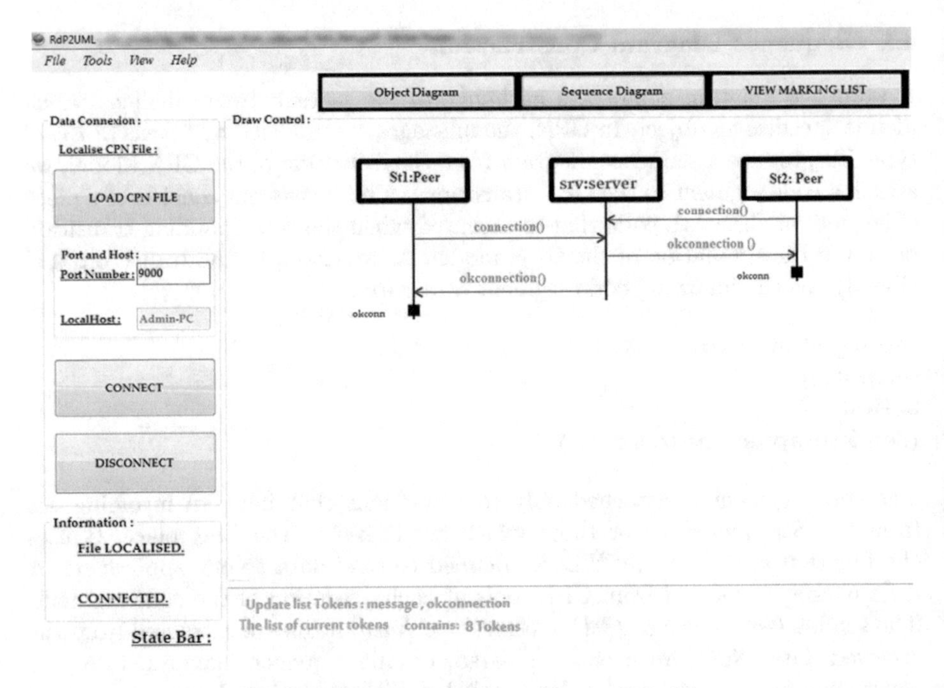

Fig. 7. Obtained sequence diagram model

communicating objects at a precise time. In CPN, a system state is given by the distribution of the tokens on the CPN places. To construct the object diagram, we proceed as follows.

We get from the CPN model all *Objet* tokens, whose form is (obj:Class, atr). Next, we produce an object for each token. The object name and attribute values are extracted from the token fields. Similarly, for each *Event* token, whose form is (src, dst, msg), we add a link between the sender object and the receiver object when a transition guarded by msg is fired. This approach is summarized by the Algorithm 2: The application of the Algorithm 2 to the example described

Algorithm 2. Algorithme ProduceObjectDiagramme(CPN)

Begin

- for each token of Objet type
  - Create an object O = obj: Class, atr
  - addObject (O) to OD ; addObject is a function that inserts an object to an object diagram
- for each token of Event type /*(src, dst, msg)*/
  - AddLink (O1,O2) ; AddLink is a function that creates a link between a source object and a target object

End

by Fig. 3, produces the object diagram shown in the Fig. 8. It represents two objects of peer class and an object of server class.

Initially the server model contains the token ("srv:server","10.0.0.1") at the place **Idle**. This token yields the object identified by (srv:Server, 10.0.0.1). When the place *Connection* of the peer model (Fig. 4) contains the token ("st1:peer", "10.0.0.10") and the place *Link* contains the token ("srv:server","st1:peer","okconnection"), the transition 't2' is fired. This firing will cause the creation of a link between the objects: (srv:Server, 10.0.0.1) and (st1:Peer, 10.0.0.10) on the object diagram by the execution of the attached code. The same procedure applies for the token ("st2:peer","10.0.0.11").

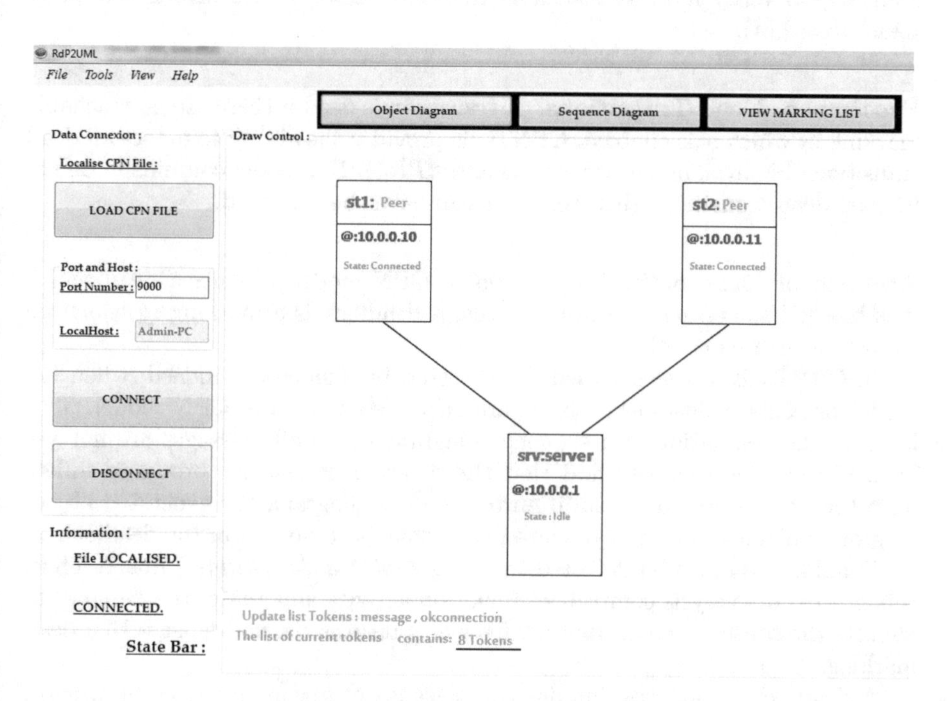

Fig. 8. Obtained Object diagram

5 Model Analysis

The verification is done by Model checking on the state space analysis. Generic and specific properties can be verified. This two points are developed in what follows.

5.1 Generic Property Analysis

This approach deals with simulation-based analysis, so, analysis result are based on simulation traces. Thus, simulation based analysis cannot provide results as strong as formal verification. In what follows, we will presents the analysis of some generic properties related to the good construction of the model, we are interested by the reachability property, presence of dead transitions and deadlock which are done on the reachability graph.

Reachability. Reachability analysis is based on the reachability graph. CPN-Tools provides the possibility to know whether the marking graph forms a strongly connected component by invoking the query AllReachable(). This query returns true when the number of strongly connected components equals to 1. In addition, to verify if M is reachable from a marking S, we invoke the query Reachable(S,M).

Presence of Dead Transitions. A transition is dead if there are no reachable marking in which it is enabled. CPNTools provides the ability to detect all dead transitions by invoking the query $ListDeadTIs()$. If a model contains in one of its part dead transitions then this part can never be activated.

Absence of Deadlocks. We say that a CPN model presents a deadlock, if it is blocked somewhere. In other words, a deadlock is a marking in which no transitions can be enabled.

In CPNTools, a deadlock can be detected by simulation. indeed, when the model presents a deadlock, tokens are still present on the CPN model. This leads to the generation of a sequence diagram where all messages are not yet transmitted. The user will find that the diagram he/she has produced differs from the one generated. In addition to sequence diagram, the generated object diagram will shows all the associated links that participated in the deadlock.

Deadlock can also be detected by using $ListDeadMarkings()$ query. Thus, this process of analysis completes simulation activity and guides the designer to identify the cause of the deadlock. This query returns the nodes list with a dead marking.

A deadlock in our case implies the presence of tokens in places $Scenario$, $Link$ or in any other places of the system at the analysis end. Indeed, after each call to $ListDeadMarkings()$ query, we get all the tokens that remain in these places and a positioning of each object on its state machine is provided to allow the user to understand the problem source. The Algorithm 3 summarizes this approach.

A system can not be formally validated only by referring to the simulation results. Simulation guides the designer to detect possible errors. The simulation combined with the generic properties analysis identifies the occurrence of the error, and consequently target the field of designer intervention.

Specific Property Verification. To validate the system faithfulness with the client requirements, specific properties are written by the designer in Computational Tree Logic (CTL), and then, verified by CPNtools.

Algorithm 3. Algorithm trace Deadlock(CPN)

Begin

1. Let L be the nodes list where a deadlock is detected by invoking the ListDeadMarkings() query.
2. **For** each node $n \in L$ **do**
   (a) Let $L2$ be the tokens list remaining in the place *Link* .
   (b) **For** each element E in the form (src, dest, msg) $\in L2$ **do**
      i. Display the message: ("$E.dst$ did not receive the $E.msg$, sent by the $E.src$").
      ii. get the place P where the $E.dst$ object is blocked.
      iii. **Returns** ($name$, $E.dst$); $name$ is the name of place P and state
   (a) - Let $L3$ be a tokens list left in the place *Scenario* .
   (b) - For each element $E2$ of form (srce, dest, msg) $\in L3$ do
      i. Display: ("The $E2.src$ object can not send the $E2.msg$ object to the $E2.dst$ object.")
      ii. Find the place P where the object $E.src$ is blocked.
      iii. get the name of place P.
      iv. Returns ($name$, $E.src$) ;
3. **If** number of tokens in place *Scenario* $= 0$ and number of tokens in place *Link* $=$ 0 **then**
4. get $L4$ list of place names that contain tokens
5. **For** each element $E \in L4$ **do**; E represents a place
   (a) let $E1$ be the list of E' tokens.
   (b) **For** each token $\in E1$
   (c) **return**($name$, $object$).

End.

The verification is done on the generated space states. The verification process can lead to two situations:

- The CTL formula is valid: we just return this result to the designer.
- The CTL formula is invalid: in this case, in order to help the designer to determine the error cause, the transitions sequence that led to the error is retrieved from the states space. Then, this sequence will be transcribed into a scenario expressed by a sequence diagram. The last transition of the counter example is put in evidence, because its causes the invalidity of the CTL formula.

The entire states space is scanned and the value n of the node where the CTL constraint condition is not fulfilled is checked. After that we have to retrieve the nodes list that lead to node n and this way compute this states space size (number of nodes), here the $NoOfNodes()$query is used. Then, the $ArcsInPath(1, n)$ query which returns the shortest path between the nodes 1 and n is invoked. In this way, the unsafe path is gotten.

The $ArcToBE(n)$ query is used to obtain informations on each transition of unsafe path. Then, we go entire unsafe path to reconstruct it in a suitable form. The Algorithm 4 summarize this approach

Once the unsafe path fixed, it will be sent to our application to extract all the tokens to retranscribe the trace into a sequence diagram. This approach is summarized by the Algorithm 5 and shown by Fig. 9.

Note: A states space is a graph where each node represents a CPN marking and each arc is labeled by the transition which caused the marking modification.

Algorithm 4. Fixing unsafe path algorithm

Input: condition of the CTL constraint (*cond*)
Output: unsafe path in string format.
begin

1. n = NoOfNodes () ; returns the size of states space G
2. **For** i from 1 to n **do**
   (a) **If** ($G(i)$, *cond* = true) **then** i ++; $G(i)$ represents the node number i in the states space.
   (b) **else** num = i (retrieve node number where *cond* is unchecked)
3. L = ArcsInpath (1, *num*).
4. **For** each E element $\in L$ **do Returns** (ArcToBE (E)).

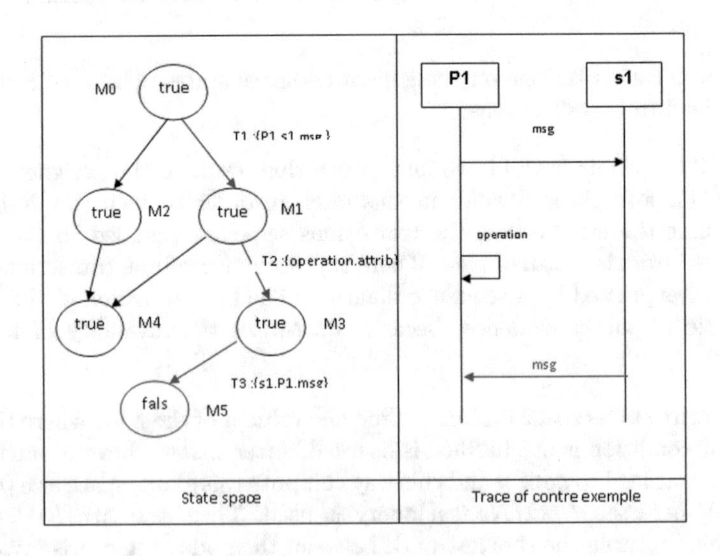

Fig. 9. Interpretation approach

Algorithm 5. interpretation algorithm (CPN model, CTL Constraint)

Input: CTL constraint and CPN model
Output: Sequence diagram.
begin

1. Generate the states space.
2. Verify CTL formula.
3. **If** the result is true **then**
4. **Returns**("Verified Constraint.")
5. else L = seek contre-example using (*Fixing unsafe path algorithm*)
6. **For** each E element $\in L$ **do**
   (a) **If** E is of the form (src, dest, msg) **then**
      i. Create an arc in the sequence diagram (src $\rightarrow$ dest)
      ii. **else if** E is of the form (src, attrib) **then**
      iii. Create an arc in the sequence diagram (src$\rightarrow$ src)

End.

6 Conclusion

Many results are communicated regarding the UML formalization. This formalization is done in order to activate some verification tasks. However, due to the difference between the UML notation and the formal language, the UML designer often do not participates in the verification process. In this paper, an approach that produces UML model from Petri Nets model analysis was proposed. The production of object diagram from Petri nets result has never been done according to our investigations. This innovative approach allows the designer to participate in the analysis precess without the need for him to know the formal model.

The Petri nets models derived from state machines are verified by model checking using CPNTools. The verification covers both the good construction of the model and its consistency with the client's requirements. The latter is undertaken by means of specific properties of the system, written by the designer in Computational Tree Logic (CTL).

References

1. Baresi, L., Morzenti, A., Motta, A., Rossi, M.: A logic-based semantics for the verification of multi-diagram UML models. ACM SIGSOFT Softw. Eng. Notes **37**(4), 1–8 (2012)
2. Bouabana-Tebibel, T.: Object dynamics formalization using object flows within UML state machines. Enterp. Model. Inf. Syst. Archit. **2**(1), 26–39 (2007)
3. Choppy, C., Klai, K., Zidani, H.: Formal verification of UML state diagrams: a Petri net based approach. ACM SIGSOFT Softw. Eng. Notes **36**(1), 1–8 (2011)

4. Hölscher, K., Ziemann, P., Gogolla, M.: On translating UML models into graph transformation systems. J. Vis. Lang. Comput. **17**(1), 78–105 (2006)
5. Jensen, K., Kristensen, L.M., Wells, L.: Coloured Petri nets and CPN tools for modelling and validation of concurrent systems. Int. J. Softw. Tools Technol. Transf. **9**(3–4), 213–254 (2007)
6. Kerkouche, E., Chaoui, A., Bourennane, E.B., Labbani, O., et al.: On the use of graph transformation in the modeling and verification of dynamic behavior in UML models. J. Softw. **5**(11), 1279–1291 (2010)
7. Kong, J., Zhang, K., Dong, J., Xu, D.: Specifying behavioral semantics of UML diagrams through graph transformations. J. Syst. Softw. **82**(2), 292–306 (2009)
8. Kuske, S., Gogolla, M., Kollmann, R., Kreowski, H.J.: An integrated semantics for UML class, object and state diagrams based on graph transformation. In: Butler, M., Petre, L., Sere, K. (eds.) Integrated Formal Methods, pp. 11–28. Springer, Heidelberg (2002)
9. Kuske, S., Gogolla, M., Kreowski, H.J., Ziemann, P.: Towards an integrated graph-based semantics for UML. Softw. Syst. Model. **8**(3), 403–422 (2009)
10. Lakos, C.: From Coloured Petri Nets to Object Petri Nets. Springer, Heidelberg (1995)
11. Mars, T.: CPN AMI. http://www-src.lip6.fr/cpn-ami
12. Pradella, M., Morzenti, A., Pietro, P.S.: Refining real-time system specifications through bounded model-and satisfiability-checking. In: 23rd IEEE/ACM International Conference on Automated Software Engineering, ASE 2008, pp. 119–127. IEEE (2008)
13. Saldhana, J.A., Shatz, S.M., Hu, Z.: Formalization of object behavior and interactions from UML models. Int. J. Softw. Eng. Knowl. Eng. **11**(06), 643–673 (2001)
14. Varpaaniemi, K., Halme, J., Hiekkanen, K., Pyssysalo, T.: PROD reference manual. Technical report B13, Helsinki University of Technology, Digital Systems Laboratory, Espoo, Finland, August 1995
15. Wang, M., Lu, L.: A transformation method from UML statechart to Petri nets. In: 2012 IEEE International Conference on Computer Science and Automation Engineering (CSAE), vol. 2, pp. 89–92. IEEE (2012)
16. Westergaard, M.: CPN tools 4: multi-formalism and extensibility. In: Colom, J.M., Desel, J. (eds.) Application and Theory of Petri Nets and Concurrency, pp. 400–409. Springer, Heidelberg (2013)
17. Yassin, A., Hassan, H.: Transformation of coloured Petri nets to UML 2 diagrams. In: Rocha, Á., Correia, A.M., Tan, F.B., Stroetmann, K.A. (eds.) New Perspectives in Information Systems and Technologies, vol. 2, pp. 131–142. Springer, Cham (2014)

Using Belief Propagation-Based Proposal Preparation for Automated Negotiation over Environmental Issues

Faezeh Eshragh[1(✉)], Mozhdeh Shahbazi[1], and Behrouz Far[2]

[1] Department of Geomatics Engineering, University of Calgary, Calgary, Canada
{feshragh,mozhdeh.shahbazi}@ucalgary.ca

[2] Department of Electrical and Computer Engineering, University of Calgary, Calgary, Canada
far@ucalgary.ca

Abstract. Automated negotiation is used as a tool for modeling human interactions with the aim of making decisions when participants have conflicting preferences. Although automated negotiation is extensively applied in different fields (e.g., e-commerce), its application in environmental studies is still unexplored. This paper focuses on negotiation in environmental resource-management projects. The primary objective of this study is to reach agreement/conclusion as fast as possible. To achieve this objective, an agent-based model with two novel characteristics is proposed. The first is automating the process of proposal offering using Markov Random Fields and belief propagation (BP). The second is the ability to estimate stakeholders' preferences through an argument handling (AH) model. The experiments demonstrated that the AH model and the BP-based proposal preparation (BPPP) approach improve the performance of the negotiation process. A combination of these two modules outperforms the conventional utility-based approach by decreasing the number of negotiation rounds up to 50%.

Keywords: Automated negotiation
Argumentation-based negotiation · Markov random field
Belief propagation · Energy-system planning

1 Introduction

Negotiation is known as one of the prominent ways of reaching an agreement when the decision makers have opposing interests (Van Kleef et al. 2006). In the past few decades, negotiation has been studied from different perspectives such as psychology (Pruitt and Carnevale 1993), economy (Kreps 1990), and computer science (Jennings et al. 2001). These studies investigate the complicated nature of a negotiation process to make it more efficient and reliable. However, the nature of the negotiation process highly depends on the context in which the negotiation occurs. While in some contexts the negotiation takes

T. Bouabana-Tebibel et al. (Eds.): IEEE IRI 2017, AISC 838, pp. 69–95, 2019.
https://doi.org/10.1007/978-3-319-98056-0_4

place between two parties (e.g., one buyer and one seller) over one single issue (e.g., the price of a product), the negotiation process involves multiple issues and multiple participants in other contexts. Environmental resource management is one of the domains that require negotiation among multiple stakeholders with different viewpoints over a variety of environmental and non-environmental issues. Particularly, negotiation can play a crucial role in the context of managing common-pool environmental resources. These resources are shared by a group of stakeholders and can be overused or congested due to poor management. An instance of such resource-management problem occurs in developing a new electricity transmission line between two locations, where each possible transmission line is a proposal over which negotiation should occur. Each proposal is itself a function of several attributes (criteria), e.g., environmental damage in different forms, development cost, influence on population, and impact on land. Past studies have shown that modeling negotiation over common-pool environmental resources using simulation tools can facilitate the decision making and management processes considerably (Ostrom et al. 1994; Bousquet and Trébuil 2005). Automated negotiation, which is facilitated by an agent-based model (ABM), is one of these tools. In this model, a set of intelligent agents interact with each other and search the space of potential agreements to find a (set of) mutually acceptable agreement(s). Automated negotiation has received considerable attention in several domains such as supply chain management (Fink 2006), political studies (Aragonés and Dellunde 2008), and e-commerce (Dworman et al. 1996; Zeng and Sycara 1998; Faratin et al., 2002; Ramchurn et al. 2006; Shojaiemehr and Rafsanjani 2018). However, only a few studies have considered modeling stakeholders' negotiations over environmental issues using ABM (Akhbari and Grigg 2013; Okumura et al. 2013; Pooyandeh and Marceau 2013, 2014; Alfonso et al. 2014). Apart from the lack of mutual collaborations between scientists in the areas of artificial intelligence (AI) and environmental studies, the other reasons for scarcity of automated negotiations over environmental issues include the high level of uncertainty in these issues, the high-stakes decisions, and the involved stakeholders who are not usually willing to reveal the details of their preferences. These factors may cause the negotiations to take too long, sometimes even without reaching an agreement. In this paper, new approaches are proposed to facilitate and accelerate automated dynamic negotiation over environmental issues. The following section provides a summary of the studies related to automated negotiation with a focus on environmental issues.

1.1 Related Work

An agent-based model (ABM) is a computational model for simulating collective behaviors in which autonomous elements, called agents, with a predetermined set of goals and actions, imitate intelligent units of the environment (Wooldridge 2001). One of the main characteristics of ABM is its autonomy as there is no global control over the agents, and they do not operate based on a globally consistent knowledge base. These agents communicate and interact with each other to exchange knowledge and achieve their goals (Oprea 2004).

ABM has been employed in various domains such as transportation management (Jennings et al. 1995), telecommunication (Bäumer and Magedanz 1999), business (Jennings et al. 1996) and electronic commerce (Cao et al. 2015). Automated negotiation is another application of ABM in which negotiation among different parties is modeled using agents and their interactions.

Three main approaches have been applied in automated negotiation: the game theory, the heuristic approach, and the argumentation-based negotiation (ABN) (Jennings et al. 2001). Game theory originates from research conducted by Neumann and Morgenstern (1945) and has its roots in economics. It provides the mathematical foundation for the analysis of interactions between self-interested agents (MacKenzie and DaSilva 2006). The heuristic approach attempts to overcome limitations of the game theory techniques by finding a satisfactory, sub-optimal solution instead of an optimal one (Kraus and Lehmann 1995; Barbuceanu and Lo 2000; An and Lesser 2012; Aydogan et al. 2013). Both game theory and heuristic approaches are proposal-based. That is, exchanging proposals is the only way through which those agents can gain information about their opponents. However, a proposal is only a point in the agreement space and contains information about neither the proposing agent nor the other parties. In this paper, the argumentation-based negotiation (ABN) approach is employed as it better fits the context of the stated problem. In ABN approaches, in addition to the proposals and reject/accept decisions, the agents can exchange supplementary information, called arguments, in support of their decisions or against their opponents' decisions (Jennings et al. 1998). That is, in addition to rejecting an offer (namely a proposal), an agent can explain which parts of the proposal are against his preferences. Different forms of arguments, rewards, treats or appeals are discussed in the literature (Ramchurn et al. 2006). One of the benefits of sharing this information is facilitating the process of estimating the space of potential agreements and, therefore, expediting the negotiation process. The agents in ABN approaches can interpret the arguments and take advantage of them through an estimation mechanism. They are, therefore, designed with more components compared with the non-ABN agents; these components are in charge of generating and interpreting the arguments (Rahwan et al. 2003). The argumentation-based approach has received significant attention within the last 20 years due to its potential in modeling real-world negotiations (Sycara 1990; Parsons and Jennings 1996; Parsons et al. 1998; Kakas and Moraitis 2006; Rahwan et al. 2004; Karunatillake et al. 2009; El-Sisi and Mousa 2012). Wang et al. applied ABN on supply chain formation where the agents use argumentation to understand the preferences of the other participants (Wang et al. 2009). In another study, ABN has been used in the design of complex infrastructure systems with many components and layers of subsystems (Marashi and Davis 2006). They proposed a systematic way to decompose the system into subsystems and sub-processes by identifying the objectives and criteria of each process and then resolving the conflicts among them using argumentation-based negotiation. ABN has also been widely employed in e-commerce (El-Sisi and Mousa 2012; Zhang et al. 2012; Jain and Dahiya 2012; Zhang et al. 2010). For example,

El-Sisi and Mousa have proposed a bargaining negotiation framework in which classic non-ABN agents are compared to ABN agents in terms of the quality of the reached agreement and number of unsuccessful negotiations (El-Sisi and Mousa 2012).

Research related to argumentation-based negotiation have mainly followed three major directions (Karunatillake et al. 2009). In the first one, called argumentation-based defeasible reasoning, arguments about different alternatives are gathered and compared with other available arguments to find possible conflicts. The argumentation system then tries to resolve the conflicts by analyzing the relationship among the arguments (e.g., support, attack, conflict, etc.) and update the agent's beliefs accordingly (Chesnevar et al. 2000; Prakken and Vreeswijk 2002; Tamani et al. 2015; Thomopoulos et al. 2015). The second approach is used in case studies where the system needs to generate and send rhetorical statements to the user. Here, arguments are characterized as structures (schema) for generating persuasive feedbacks (Gilbert et al. 2004). The third group of studies in this area employ arguments as a model of interaction when there are conflicts between negotiating parties. The arguments in this approach are generated under the structure of dialogue games using a common communication language and based on a set of pre-defined rules governing the negotiation (Karunatillake et al. 2009). The argument handling in the current study follows the third direction as the agents communicate and generate arguments using a dialogue protocol, a set of negotiation rules and a communication language. The formal analysis of the relationships among arguments is therefore out of the scope of this paper. That is, arguments are represented as phrases justifying the agents' decisions for rejecting/accepting an offer and all the agents' arguments will be considered in advancing the negotiation regardless of possible conflicts among them. It is also assumed that the negotiation parties only submit reasonable and sensible arguments.

The order of offered proposals is another critical matter in the negotiation process that needs to receive more attention. The order by which the proposals are selected and offered to negotiation parties may dramatically influence the pace of the negotiation process and the quality of its outcomes. This is specifically important when the space of potential agreements is large and high-dimensional. However, for the proposing agent, determining the right proposal based on his limited knowledge of the preferences of the other parties is a challenging task. In the current study, we applied a belief propagation-based approach to tackle this problem. Belief propagation-based techniques have rarely been used in automated negotiation. In Robu et al. (2005) and Hadfi and Ito (2015, 2016), a belief propagation-based technique is employed to represent the utility space and facilitate its exploration. The focus of these studies is the optimal solution for one participant using his utility graph and constraints. However, such an approach is not applicable to environmental negotiations, in which many participants with multiple perspectives are involved, and all their preferences should be considered simultaneously.

1.2 Objectives and Contributions

In this paper, we present an automated negotiation solution for modeling negotiations over multi-criteria, multi-participant decision-making problems, such as the ones that happen in environmental resource management. The main objective of this solution is to reach agreement/conclusion in as few rounds of negotiation as possible. This work is an extension of our previous conference presentation (Eshragh et al. 2017).

The proposed solution is implemented as an ABM with several processing modules and knowledge bases. The novel characteristics of this ABM are twofold: further automation and intelligence. The first novelty is based on automating the proposal-offering process. This problem is graphically modeled as hidden Markov Random Fields (MRF), where attributes of the proposals are considered as observable nodes, and the possible values (states) of the attributes are the hidden variables. A joint probabilistic model, which depends on preferences of negotiation parties, is built over the attributes and hidden variables. The direct statistical dependencies between hidden variables are modeled by connecting hidden variables via undirected edges in the graph. A pair of states are connected if they can simultaneously be used to form a proposal, and the length of this edge describes their consistency. Inference on this graphical model is performed using min-sum loopy belief propagation. Having all available information in one MRF model, the best offer is selected in each round of negotiation. As the negotiation proceeds, the model is dynamically rebuilt based on re-estimating the preferences of negotiation parties, which is performed by the second contribution of this paper. To this end, an argument handling (AH) module is developed, using which the agents exchange information with the proposer agent in each round of negotiation. The module applies the exchanged information to update the knowledge of the proposer agent about the preferences of the other parties. The proposer agent, then, uses these new preference estimations to rebuild the MRF model and prepare a more mutually acceptable proposal. The dynamic nature of the MRF in combination with the AH model makes the system a very suitable match for the target problem. The proposed method is successfully employed in the case study of energy-system planning. We have also evaluated this method in a real-estate case study to verify the performance when dealing with larger problems regarding the search space and the dimensionality of the involved criteria.

The rest of the paper is organized as follows. In Sect. 2, the components of automated negotiation are described. Two case studies, energy-system planning in Alberta and King County House Sales, as well as experiments, implementation details, and the results are discussed in Sect. 3. The conclusions are presented in Sect. 4.

2 Automated Negotiation

Automated negotiation consists of three main components: negotiation protocol, negotiation object, and negotiation strategy (Lomuscio et al. 2001).

The negotiation protocol defines the set of rules governing the interactions between agents. It determines the possible types of participants, the negotiation phases, the negotiation rules, and the possible actions for each participant in each phase. In this study we have two types of agents. The first one is the proposer agent, representing the party in charge of preparing and offering proposals in each round of negotiation. The second type of negotiating agents represents other stakeholders who receive the proposals and decide whether to agree or disagree with them. In accordance with most real-world negotiations, it is assumed that the proposer agent offers only one proposal in each round of negotiation and conceals his knowledge of other possible offers from other agents. The agents follow a simple protocol, shown in Fig. 1. Rectangles in this figure represent the set of actions and the lines between them show the order of possible moves between these actions. The proposer initiates the negotiation process by sending an offer to the stakeholder agent. The proposed offer will be either accepted or rejected by the agent. In case of rejection, the proposer will ask for the reason (Challenge) and receives an argument (or a set of arguments) in response (Assert). The process continues till the stakeholder accepts the offered proposal or the proposer calls for a withdrawal of the process. The call for withdrawal happens when the proposer runs out of proposals (i.e. all possible proposals are already offered and rejected by the stakeholders). Here, we assume all agents are rational.

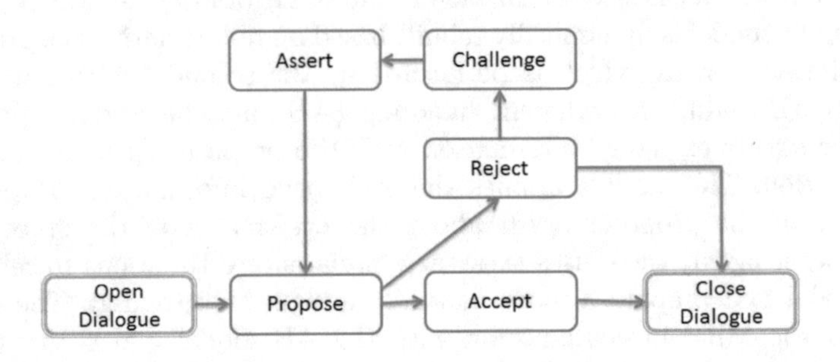

Fig. 1. Negotiation protocol

The second component of automated negotiation, called negotiation object, corresponds to the range of criteria over which the negotiation happens. The proposed methodology handles multi-criteria negotiations. In each round of negotiation, the proposer assigns values to this set of criteria which all together form a proposal. The stakeholders have to decide whether they accept this proposal or reject it. In this study, we assume that the criteria (attributes) can be numerically quantified. The third component is the agents' decision-making process, also called agents' strategy, used by the agents to act according to the negotiation protocol. It is basically the agent's plan for achieving its goals (Lomuscio et al. 2001). While the negotiation protocol is public and available to all participants, the agent's strategy is always private. In this study, the proposer agent's strategy is more about the way it prepares a proposal while the stakeholder agents' strategies explain how they evaluate a received proposal and generate arguments about their decisions. In the next sections, these strategies will be explained in more details.

In Sect. 2.1, the ABM and its components are presented. In Sect. 2.2, the problem of automated negotiation is broken down and formulated. Sections 2.3 and 2.4 describe the BPPP approach and the statistical solution to determine the final proposal in each round of the negotiation. Section 2.5 presents the details of argument handling model.

2.1 Agent-Based Model

In this study, an ABM has been employed to model a one-to-many type of negotiation among stakeholders with different perspectives. Each agent in the model has preferences over certain criteria that are designed based on the stakeholder's concerns and interests. These stakeholders represent the groups or individuals who are either involved in the decision-making process or affected by the final decision (Freeman 2010).

Each agent in the ABM has access to a set of private and public knowledge bases containing information about the agent's goals, beliefs, and preferences. This information is gathered either from public documents and datasets or directly from the stakeholders. A part of the agents' knowledge that relates to the external environment and other participants may change during the negotiation as a result of re-estimation processes. In the proposed model, the re-estimation process happens in the argument handling model, explained in Sect. 2.5. The proposer agent has a proposal database of all possible alternatives as well as a database for previously offered proposals. These records can be used for keeping track of negotiation rounds and future references (Rahwan et al. 2004). All agents have databases of their preferences and criteria. This database is utilized in assessing alternatives and generating arguments during the negotiation process.

A schematic representation of the negotiation system developed in this study is displayed in Fig. 2. The system includes an ABM as well as the databases and the external modules required for generating proposals, selecting among them, and handling the arguments.

2.2 Problem Statement

Consider a set of $x+1$ stakeholders, $S = \{s_p, s_1, s_2, \cdots, s_x\}$, who are negotiating over a set of proposals, $P = \{p_1, p_2, \cdots, p_y\}$. Here, s_p represents the proposer agent who initiates the negotiation and proposes a new offer in each round. Each proposal $p_j (j = 1, \cdots, y)$ is described by a set of attributes $A = \{a_1, a_2, \cdots, a_z\}$; that is, proposal p_j is a unique combination of the values of these attributes as $A_j = \left\{a_1^j, a_2^j, \cdots, a_z^j\right\}$. The preference thresholds of stakeholder $s_i (i = 1, \cdots, x)$ over attribute $a_k (k = 1, \cdots, z)$ is represented as his minimum and maximum acceptable values for that attribute, denoted by Min_{a_k, s_i} and Max_{a_k, s_i}. The

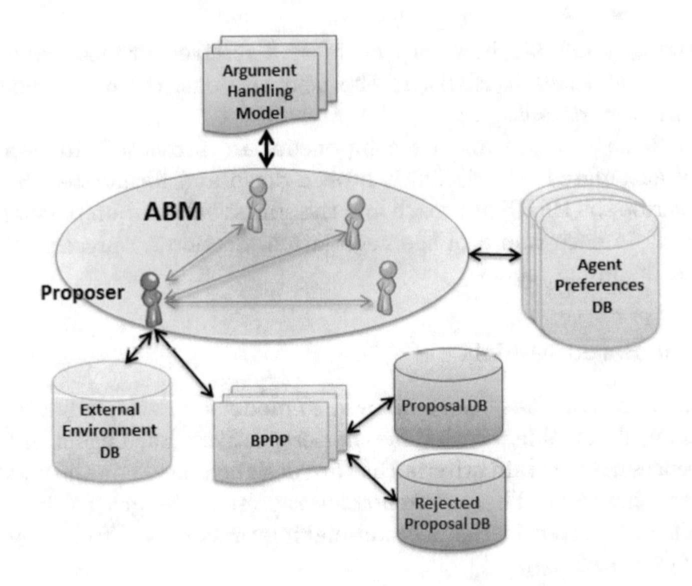

Fig. 2. A schematic representation of the proposed negotiation system

agents negotiate to find a proposal, if any, that meets all agents' preferences. Without loss of generality, we assumed that all agents (e.g., stakeholders) have the same importance in the negotiation process. The negotiation ends when either all agents agree on a proposal, or the proposer confirms that there is no more proposal to offer and terminates the negotiation without reaching an agreement.

2.3 Belief Propagation-Based Proposal Preparation (BPPP)

In each round of negotiation, a proposal is selected by the proposer agent to be offered to the stakeholders. The goal of the proposer is to find a proposal that is aligned with his preferences and has the highest probability of being accepted by other agents. In the beginning, the proposer agent knows nothing about the stakeholders' preferences; therefore, it acts based on its preferences as well as some assumptions about the ranges of acceptable values by other agents. However, as the negotiation proceeds and the feedback (e.g., agent's decisions and arguments) are received, he gradually learns about other participants' preferences. He will then utilize the recently learned knowledge about a particular stakeholder in finding a potential agreement with that stakeholder. From the probabilistic point of view, this proposal preparation process can be modeled as an inference problem; That is, the proposer infers the most probable proposal given the likelihood (based on the preferences of a specific stakeholder over each attribute of the proposal) and the prior (based on the interdependence of specific attributes). In the current study, the proposal preparation problem is modeled using Markov Random Fields (MRF) where belief propagation is applied to approximate the inference.

In the MRF, for each stakeholder s_i an undirected graph is defined with attributes $A = \{a_1, a_2, \cdots, a_z\}$ as its nodes. Each attribute $a_k (k = 1, \cdots, z)$ has a set of alternate states (values) $V_{a_k} = \{v_1^k, v_2^k, \cdots, v_{n_k}^k\}$. The problem is finding the optimal state for each attribute and then identifying the most optimal proposal among all possible alternatives $P = \{p_1, p_2, \cdots, p_y\}$ based on the selected states. To figure this problem out, the proposer needs to know the preferences of the stakeholder s_i, namely $\{Min_{a_k,s_i}, Max_{a_k,s_i}\}$. However, in real-world situations, stakeholders usually do not share this sort of information, and it is often considered private. Therefore, the proposer agent starts with an estimation of these thresholds and tries to learn them through the feedback he receives from others.

To solve this optimization problem, an energy minimization framework can be employed where a global energy function penalizes each alternative for being either dissatisfactory to the stakeholder (unary cost) or not consistent with other attributes' values (binary cost). The dissatisfaction level of a state of an attribute for stakeholder s_i is defined as the unary cost of that state. In mathematical terms, the cost of assigning state $v_l^k (l = 1, \cdots, n_k)$ of node $a_k (k = 1, \cdots, z)$ in negotiation with stakeholder s_i, the unary cost $U_{l,i}^k$, is defined as follows.

$$UC_{s_i}(v_l^k) = U_{l,i}^k = (v_l^k - Min_{a_k,s_i})/(Max_{a_k,s_i} - Min_{a_k,s_i}) \tag{1}$$

This reverse-scoring function is selected to reflect the fact that the further the cost of an alternative is from the stakeholder's preferences, the lower the chance of its selection.

The binary cost is defined to enforce the compatibility between two states of two different attributes. It enforces the fact that two states of any two attributes should belong to the same proposal for them to appear together. Quantitatively, the binary cost of the combination of the state $v_{l_1}^{k_1}$ of node a_{k_1} and state $v_{l_2}^{k_2}$ of node $a_{k_2} (k_1 = 1, \cdots, z$ and $k_2 = 1, \cdots, z)$, denoted as $B_{l_1,l_2}^{k_1,k_2}$, is calculated as below.

$$B_{l_1,l_2}^{k_1,k_2} = \begin{cases} \infty, & \text{if } P_1 \cap P_2 = \emptyset \\ \frac{\sum_{j|p_j \in P_1 \cap P_2} c_j}{|P_1 \cap P_2|}, & \text{if } P_1 \cap P_2 \neq \emptyset \end{cases} \tag{2}$$

In Eq. 2, $P_1 \subset P$ is a sub-set of all the proposals that are compliant with state $v_{l_1}^{k_1}$; i.e., $P_1 = \left\{p_j | a_{k_1}^j = v_{l_1}^{k_1}\right\}$. Similarly, $P_2 \subset P$ is a sub-set of all the proposals that are compliant with state $v_{l_2}^{k_2}$; i.e., $P_2 = \left\{p_j | a_{k_2}^j = v_{l_2}^{k_2}\right\}$. Also, the variable c_j refers to the cost of the proposal p_j according to the proposer himself.

Having the unary and binary costs, a belief propagation-based approach is used to find the solution with the minimum energy. Here, the belief of node $a_{k_1} (k_1 = 1, \cdots, z)$ about its state $v_{l_1}^{k_1} (l_1 = 1, \cdots, n_{k_1})$ during the negotiation with stakeholder $s_i (i = 1, \cdots, x)$ is defined as,

$$bel_{k_1}(v_{l_1}^{k_1}) = U_{l_1,i}^{k_1} + \sum_{a_{k_t} \in N_{k_1}} msg_{k_t \to k_1}(v_{l_1}^{k_1}) \tag{3}$$

where $U_{l_1,i}^{k_1}$ is the unary cost of state $v_{l_1}^{k_1}$ for node a_{k_1}; the set N_{k_1} includes all the nodes connected to a_{k_1}; and, $msg_{k_t \to k_1}(v_{l_1}^{k_1})$ is the message that node a_{k_1} receives from the node a_{k_t} about the state $v_{l_1}^{k_1}$, which is defined as follows.

$$\begin{aligned} msg_{k_t \to k_1}(v_{l_1}^{k_1}) = min_{l_t=1,..,n_{k_t}}(U_{l_t,i}^{k_t} + B_{l_1,l_t}^{k_1,k_t} \\ + \sum_{a_{k_u} \in \{N_{k_t} \setminus a_{k_1}\}} msg_{k_u \to k_t}(v_{l_t}^{k_t})) \end{aligned} \tag{4}$$

Here, $U_{l_t,i}^{k_t}$ is the unary cost of state $v_{l_t}^{k_t}$ for node a_{k_t}; $B_{l_1,l_t}^{k_1,k_t}$ is the binary cost of the state $v_{l_1}^{k_1}$ of node a_{k_1} and state $v_{l_t}^{k_t}$ of node a_{k_t}; and, the set $\{N_{k_t} \setminus a_{k_1}\}$ includes all the nodes connected to a_{k_t} except a_{k_1}.

The nodes pass messages to each other for several iterations till we ensure all information has been transmitted through messages, and the beliefs of all the nodes about their states reach a steady state. In this stage, the solution is the set of the attribute states with the lowest belief values.

2.4 Z-Scoring Approach for Selecting Among the Prepared Proposals

The proposer agent builds one MRF model, explained in Sect. 2.3, for each stakeholder involved in the negotiation process to find a proposal with the highest acceptance likelihood for that particular stakeholder. By the end of the BPPP processes, the proposer has x-number of proposals (each prepared for a stakeholder). Therefore, he needs to choose one of these proposals as the final one to offer. In this study, we applied z-scoring to facilitate the selection among these proposals. A z-score indicates how many standard deviations a score is from the mean. For the score γ, the z-score, shown as ζ, is calculated as below,

$$\zeta = \frac{\gamma - \mu}{\sigma} \tag{5}$$

where γ and σ are the mean and standard deviation of the population, respectively. This is the way that a z-score is calculated for the values of one attribute. Here, however, we are dealing with proposals, which usually are consisted of more than one attribute. To calculate the z-score for each alternate proposal in our database, the process shown in Fig. 3 is used.

In the first step, the z-score for each attribute is calculated. The first requirement for analyzing the z-score is calculating the mean and the standard deviation of each attribute over all the proposals. Considering μ_{a_k} and σ_{a_k} as the mean and standard deviation of attribute a_k, the z-score of a_k^j (the value of attribute a_k in proposal p_j) is calculated as follows.

$$\zeta_k^j = \frac{a_k^j - \mu_{a_k}}{\sigma_{a_k}} \tag{6}$$

The next step is calculating the mean and standard deviation of the z-scores of all attributes for each proposal. For proposal p_j, these values are calculated as in Eqs. 7 and 8,

$$\mu_j = \frac{\sum_{k=1}^{z} \zeta_k^j}{z} \tag{7}$$

$$\sigma_j = \sqrt{\frac{\sum_{k=1}^{z} \left(\zeta_k^j - \mu_j\right)^2}{z}} \tag{8}$$

where z is the total number of attributes.

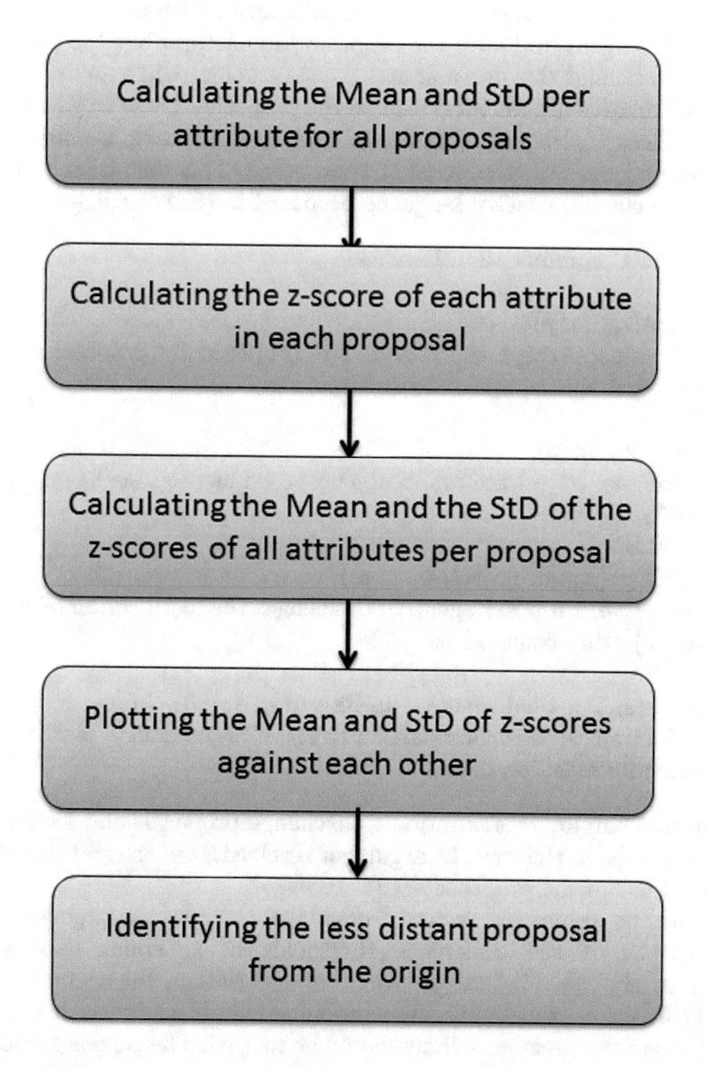

Fig. 3. The method of calculating the z-score for each alternative proposal

In the final step, each proposal p_j is represented by a point in the 2D Euclidean space, with Cartesian coordinates of (μ_j, σ_j). The proposal with the least distance to the origin will be then selected and offered by the proposer agent to the other agents in the current round of negotiation.

2.5 Argument Handling Model

Once a proposal is offered, stakeholder agents start to evaluate it based on their preferences. Based on the results of these evaluations, the agents send feedback to the proposer. If the offered proposal meets all the preferences of a stakeholder, it will be accepted by that particular stakeholder agent with no further comments. However, if the agent evaluates the proposal as unacceptable, the proposer asks for the reason behind this decision and the stakeholder will support his decision by a (set of) relative argument(s) sent to the proposer. These steps are shown in Fig. 1 with 'Accept', 'Reject', 'Challenge' and 'Assert' boxes. Definition 1 represents a more formal representation of these steps. This definition is a modified version of the communication language proposed in (Karunatillake et al. 2009).

Definition 1. Communication Language
ACCEPT
Usage: Accept(s_p, s_i, p).
Informal Meaning: Accept the proposal p, proposed by proposer agent s_p to stakeholder agent s_i
REJECT
Usage: Reject(s_i, s_p, p).
Informal Meaning: Reject the proposal p proposed by proposer agent s_p to stakeholder agent s_i
CHALLENGE
Usage: Challenge(s_p, s_i, Reject(s_i,s_p, p))
Informal Meaning: Proposer agent s_p Challenges the justification of stakeholder agent s_i for rejecting proposal p.
ASSERT
Usage: Assert(s_i, s_p, Challenge(s_p, s_i, Reject(s_i, s_p, p)), λ)
Informal Meaning: s_i Asserts a particular set of arguments λ to s_p in response to challenging its rejection decision.

The agents employ this notation to exchange proposals and feedbacks. Our focus, however, is on the way the argument received from the stakeholder agents affect the beliefs of the proposer agent.

Based on the argument received from stakeholder s_i, the proposer improves his estimations of the preference thresholds of s_i about each attribute, $\{Min_{a_k,s_i}, Max_{a_k,s_i}\}$. If at the tth round of negotiation, the argument received from stakeholder s_i states that the value of attribute a_k is too high, then the upper-bound of the preference limit should be revised. The current upper-bound, $Max^t_{a_k,s_i}$, is then re-estimated to $Max^{t+1}_{a_k,s_i}$ as below,

$$Max_{a_k,s_i}^{t+1} = \begin{cases} a_k^j, & \text{if } Max_{a_k,s_i}^{t} > a_k^j \\ \\ Max_{a_k,s_i}^{t} - (\eta \times Max_{a_k,s_i}^{t}), & \text{if } Mux_{a_k,s_i}^{t} \leq a_k^j \end{cases} \tag{9}$$

where a_k^j is the value of attribute a_k in the current proposal p_j and η represents the ratio by which the upper-bound of the preference limit is updated.

On the other hand, if the feedback received from stakeholder s_i argues that the value of attribute a_k is too low, then the lower-bound of the preference limit should be readjusted. The current lower-bound, Min_{a_k,s_i}^{t}, is then updated to Min_{a_k,s_i}^{t+1} using the following equation.

$$Min_{a_k,s_i}^{t+1} = \begin{cases} a_k^j, & \text{if } Min_{a_k,s_i}^{t} \leq a_k^j \\ \\ Min_{a_k,s_i}^{t} + (\eta \times Min_{a_k,s_i}^{t}), & \text{if } Min_{a_k,s_i}^{t} > a_k^j \end{cases} \tag{10}$$

Here, η represents the ratio by which the lower-bound of the preference limit is re-estimated.

The new preference limits for attribute a_k according to stakeholder s_i will be then used to update the unary cost of its lth state by substituting the preference thresholds in Eq. 1 with the new preference limits from Eqs. 9 and 10. After updating the unary costs based on the received arguments, the BPPP approach will be used again to find the most appropriate proposal for the next round of negotiation. The process will continue till either all agents agree on a proposal or they terminate the negotiation with no agreement.

3 Experiments

We applied the proposed methodology in two different case studies; the first case, which discusses the energy-system planning in Alberta, deals with a set of GIS data, gathered from Alberta Biodiversity Monitoring Institute (ABMI) and Alberta Environment and Parks (AEP) public resources. The second case contains real estate data from King County, US, between May 2014 and May 2015. Although this case study is not related to environmental issues, it involves multi-criteria, multi-participant negotiations and gives us the opportunity to assess the performance and scalability of our proposed method on a problem with a larger solution space. The dataset includes 21614 records (comparing to 100 alternatives in the case study related to energy-system planning) and higher data dimensionality (eight attributes for each proposal comparing to five attributes in the case study related to energy-system planning).

In these experiments, the performance of a negotiation method is measured as the number of rounds within which the negotiation process terminates. The termination happens either when participants reach a mutually acceptable agreement or when searching for the possible agreement stops (e.g., due to time restriction or finding no agreement among the participants) (Jennings et al. 2001).

To examine the performance of the proposed techniques, a set of experiments has been designed. These experiments evaluate the effect of the AH model and BPPP approach on the performance of the negotiation process. We have also analyzed different settings in the BPPP approach to find the optimal intervals for attribute-value discretization in the first case study. To compare the performance of the proposed negotiation strategy with the utility-based approach, we have conducted another set of experiments for both case studies. The utility-based approach is one of the most popular methods of evaluating proposals and is usually simplified to the weighted average of attribute values. That is, the utility of proposal p_j for stakeholder s_i is calculated as:

$$Utility_i(p_j) = \sum_{k=1}^{z}(w_k^i \times a_k^j) \tag{11}$$

where w_k^i is the weight of attribute a_k according to stakeholder s_i and a_k^j is the value of attribute a_k in proposal p_j.

The rest of this section explains the case studies and the conducted experiments for each case study.

3.1 Case Study A: Energy-System Planning in Alberta

During the past decade, the electricity demand in Alberta has arisen, and more reliable electricity grids need to be developed to transfer the generated power to the consumers. Besides selecting among available technologies (e.g., transmission lines and substations), finding the most reasonable routing option (to link the supply source and the demand center) is a key problem in these sorts of projects where both environmental and non-environmental factors are involved. In the first case study of this research, we investigated an electricity transmission project, in which the supply source is a hydropower plant near Slave River and Forth Smith city at the border of Alberta and Northwest Territories. It is shown as a red star on the map of Fig. 4. "Thickwood Hills 951s" and "Ells River 2079s" are the substations nominated to receive the transferred electricity. Orange triangles illustrate these two substations in Fig. 4. There are many alternative routes to connect the hydropower plant and the proposed substations, and the goal is to find the one that satisfies every stakeholder involved in the project. To achieve this goal, a set of criteria has to be considered, e.g., the area, type, and the coverage of the land that will be affected, the development and construction costs, the environmental impacts (e.g., wildlife), and the population that will be influenced by the route.

In our study, three categories of stakeholders are considered, including: first nations, industrial parties, and environmentally-focused groups (Table 1). Each category has preferences over a set of issues (attributes) based on their primary interests and concerns. For example, stakeholders in the environmentally-focused category are mostly concerned about specific issues including the disturbed area of forests, wetlands and wildlife, and, therefore, their preferences on these issues will influence their decisions during the negotiation process.

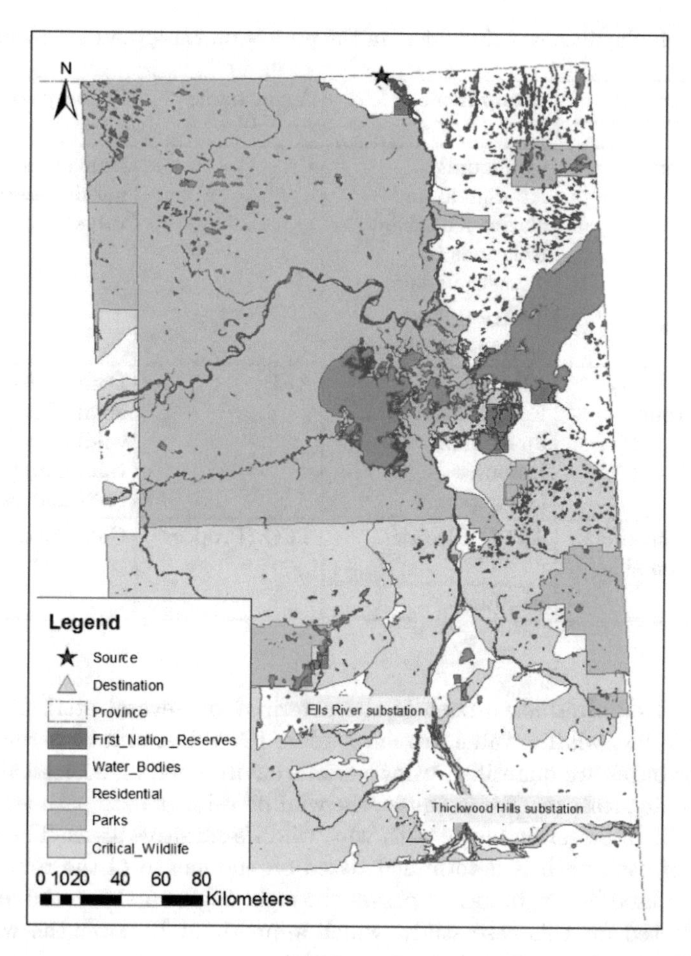

Fig. 4. The study area located near the Slave River and Forth Smith city at the border of Alberta and the Northwest Territories

3.2 Data Preparation and Implementation

The developed ABM is implemented using thread processing in Java. The agents have some private and public knowledge bases. Some of these databases are developed using Microsoft SQL 2010, and the rest are shared files accessible to all the stakeholders.

Specifying the Search Space

Several GIS data layers such as the maps of forests, lakes, rivers, caribou zones, roads, and first nations' reserves have been acquired through Alberta Biodiversity Monitoring Institute (ABMI) and Alberta Environment and Parks (AEP) public resources. A set of alternative routes from the supply source to the destination substations have been determined using a python script. This script uses Arcpy library Least Cost Path (LCP) analysis to find the alternative

Table 1. Significant stakeholders in the project on energy-system planning

Stakeholder category	Group name	Agent name	Primary concerns
First Nations	Community, Aboriginal and Native American Relations in TransCanada	FN	Damage to first nation reserves (FN value)
	Treaty 8 First Nations of Alberta		
Environmentally focused groups	Alberta Environment and Sustainable Resource Development	AEP	Damage to forest areas (Forest value), wildlife (Wildlife value), and wetlands (Wetland value)
Industries on the transmission side	ATCO Electric	TFO (Proposer)	Construction costs
	AltaLink		

solutions. Each alternative path is characterized by several attributes including forest value, wildlife value, wetland value, FN value, and construction cost. These attributes are quantified using various environmental, ecological, cultural and economic measures. For example, the wildlife value of each path is calculated based on the intersection of the path with wildlife-sensitive areas. The construction cost of each path is determined based on the length of the route and the type of the land/water bodies it passes through. Figure 5 shows the alternative routes selected for this case study, which approximately cover the whole area between the source and the destination points.

Discretizing Attribute Values

Since the attributes are quantified with continuous values, the number of possible states for each attribute would be infinite. Therefore, to make the problem solvable, the continuous values of the attributes should be reduced to finite discrete states.

The optimal number of discrete states should minimize the number of negotiation rounds. To determine this optimal value, we have empirically evaluated the influence of five different numbers of states (5, 10, 15, 20 and 25 states for each attribute) on the number of negotiation rounds with and without the AH model. As it is shown in Fig. 6, the case with 15 states for each attribute results in the minimum number of rounds for both settings (i.e., with/without argument handling module). Increasing the number of states beyond 15 results in degraded performance; using very small intervals to discretize attribute values leads to an increased number of negotiation rounds due to increasing the number of inter-state switches that the proposer should perform to find a suitable

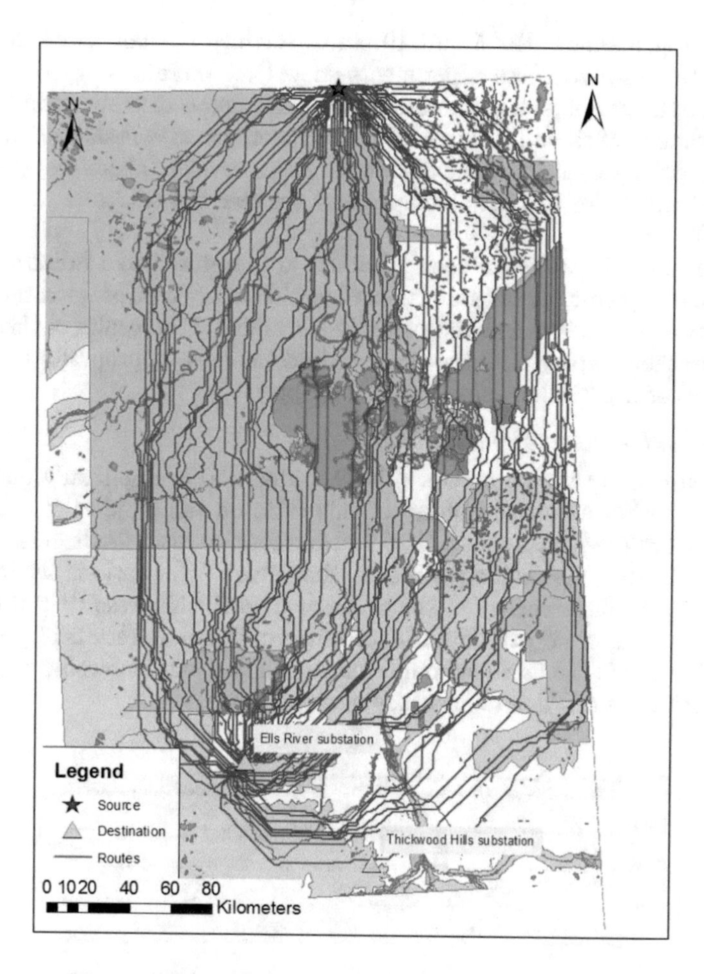

Fig. 5. Selected routes in the data preparation phase

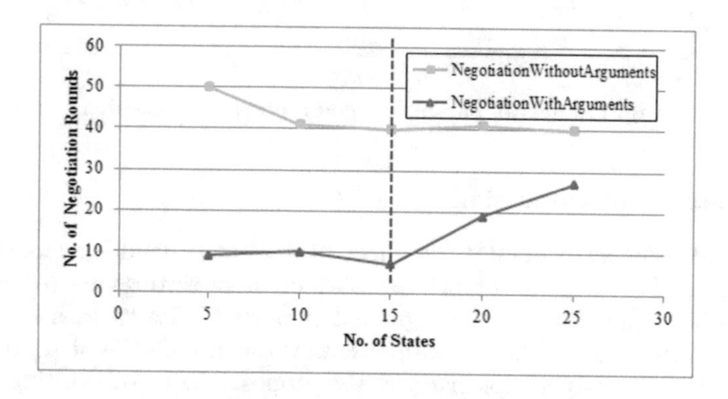

Fig. 6. Effect of number of Ranges on rounds of negotiation

state. As Fig. 6 shows, the 5 and 10 states settings are not as efficient as 15 states setting either. These experiments show that there is a tradeoff between the size of the search space (increasing by using small discretization intervals) and the chance of reaching states that have overlaps with stakeholders' preferences (increasing by using large discretization intervals). Therefore, setting the number of the states to 15 is a compromise between the extremes (5 states and 25 states).

As a result, the values of each attribute are clustered to 15 discrete states. Then, the unary cost of each state, as well as the binary cost of the combinations of every two states, is calculated using Eqs. 1 and 2. The results of these calculations are then employed by the ABM to find the most appropriate proposal in each round of negotiation using the BPPP approach.

Determining the Value of the Parameter η

For this case-study the parameter η in Eqs. 9 and 10 is empirically determined through experiments on the effects of this ratio on the number of negotiation rounds. Figure 7 indicates that 0.05, 0.1 values for this ratio result in less number of rounds. In the following experiments, the value of this ratio is set to 0.05 as it causes the smallest change to the preference limit while resulting in the least number of rounds of negotiation. Higher values of this ratio results in passing the threshold limits in the first few rounds of negotiation and increasing the number of negotiation rounds due to accuracy of estimations.

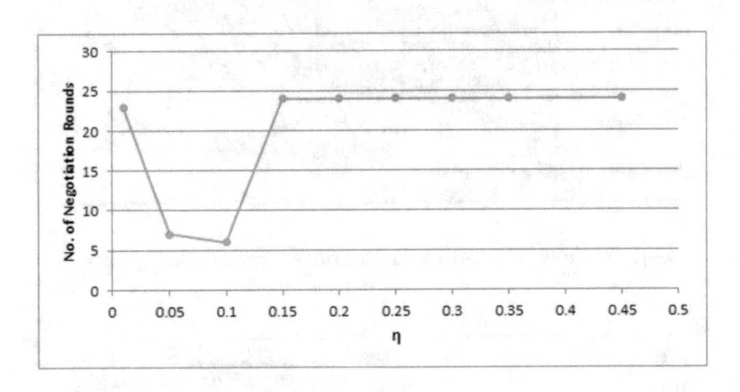

Fig. 7. Effect of parameter η on rounds of negotiation

3.3 Results and Discussions

The first experiment examines the effect of argument handling model on the performance of the negotiation process. Two different settings are tested in this experiment. In the first one, no argument is passed between each stakeholder and the proposer agent. The only response that the proposer receives from other parties is whether they accept or reject the proposal. With this setting, it takes 40 rounds of negotiation before the agents reach an agreement. In the second setting, the stakeholders provide arguments to justify their responses. In this case, the number of the negotiation rounds decreases to 7 rounds. To have a better idea of the simulation time, the first experiment (i.e. without AH module) with 40 negotiation rounds takes about 0.425 s and the second one (i.e. with AH module) with 7 rounds takes around 0.15 s. About 40% of the running time is spent in BPPP module, which is the most time consuming element of the system. Note that the experiments were performed on a quad-core machine (Intel Core i7-860, 2.80 GHz, 8 GB RAM, Windows 7).

Figure 8 shows the attribute values of the offered proposal in each round of negotiation before reaching an agreement with all the stakeholders. The attribute values of the offered proposals in the first and the second settings are illustrated by blue and red lines, respectively. In each sub-plot of Fig. 8, the green line represents the maximum value of the attribute which can be accepted by the relevant agent. For instance, the sign of reaching an agreement with the FN agent is that the "FN value" of the offered proposal becomes less than 1000 units. It should be noted that this value and other thresholds are completely hidden from the proposer agent.

We have also conducted a set of experiments to examine how the BPPP approach affects the negotiation process. To this end, we introduced a disagreement distance measure for each proposal offered to the stakeholders. This measure is defined as the maximum of the average distances of normalized attribute values in the proposal to the normalized preference thresholds of the stakeholder. For proposal p_j, the disagreement distance measure is calculated as below.

$$\Delta(p_j) = Max_{i=1}^{x}\left(\frac{\sum_{k=1}^{z}\left(\left|a_k^j - T_{a_k,s_i}\right|\right)}{z}\right) \tag{12}$$

In Eq. (12), a_k^j is the normalized value of attribute a_k in proposal p_j and T_{a_k,s_i} is the normalized threshold on attribute a_k according to the preference of stakeholder s_i. The distance $\Delta(p_j)$ measures the level of disagreement between the proposer and the stakeholders upon offering the proposal p_j; the less the value of $\Delta(p_j)$, the higher the level of agreement. Accordingly, the negotiation terminates when $\Delta(p_j)$ reaches zero. Figure 9 represents the disagreement distances when we run the negotiations with and without using the BPPP approach. This set of experiments has been conducted without using the AH model.

As illustrated in Fig. 9, without utilizing the BPPP approach, the disagreement distances fluctuate considerably. However, the BPPP approach reduces the number of fluctuations to a great extent. This is because using the BPPP approach helps the proposer agent to take other stakeholders' preferences into account and therefore, reduces the disagreements to a great extent. It is shown in Fig. 9 that even without arguments, the BPPP approach can speed up the negotiation process up to approximately 1.5 times. The green and red dashed lines show where the negotiation ends with and without BPPP, respectively.

In another set of experiments the BPPP approach has been compared with the utility-based approach. Figure 10 represents the results of these experiments with and without arguments. As illustrated in Fig. 10, the BPPP approach accelerates the negotiation process regardless of using the AH model. It is also

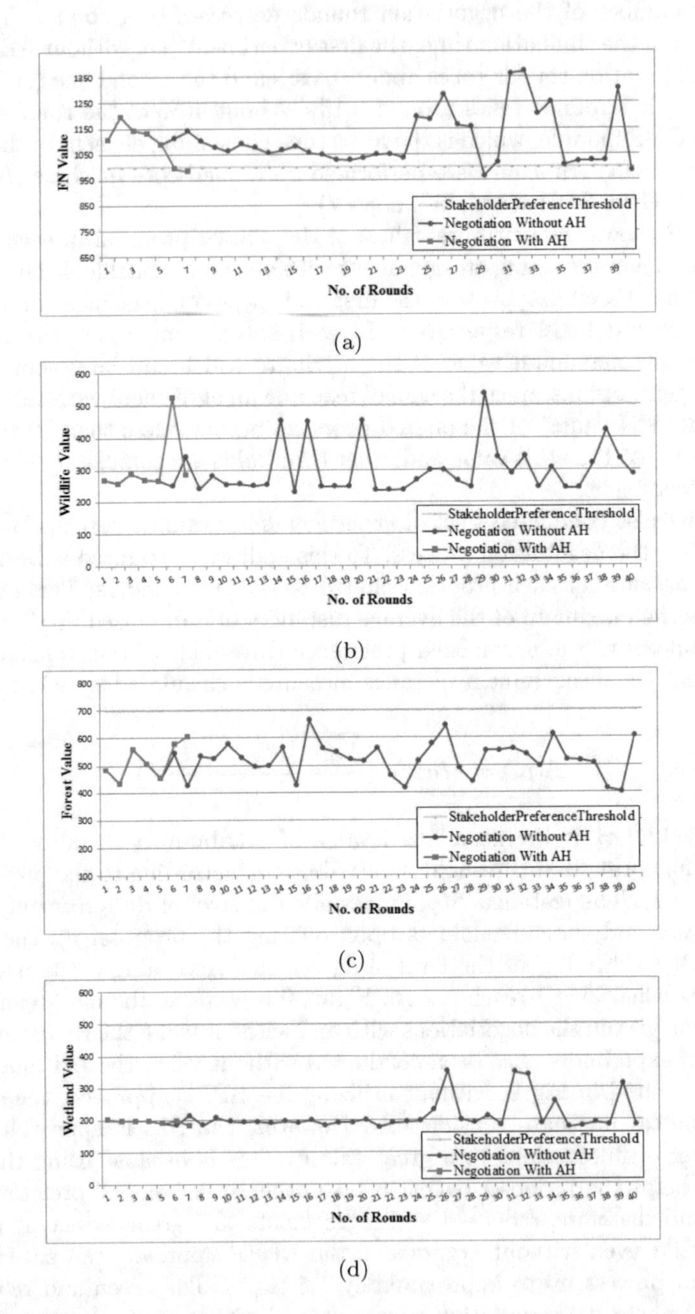

Fig. 8. The attribute values of the offered proposals in two different settings - the case of 15 states; (a) First-nation values; (b) wildlife values; (c) forest values; (d) wetland values

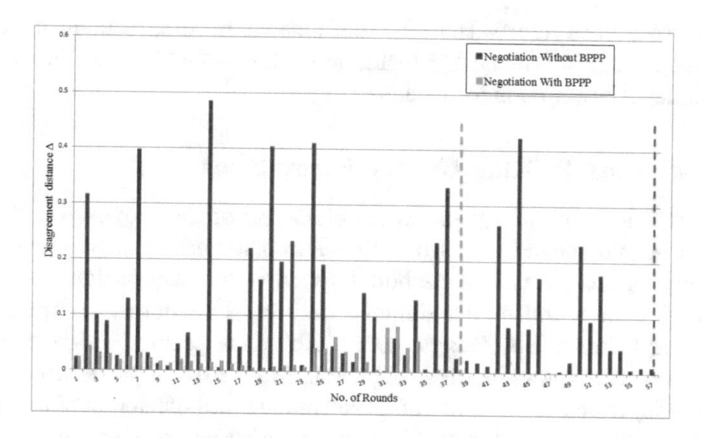

Fig. 9. Disagreement distances in negotiation over the first case study with and without using the BPPP approach. The dashed lines represent the end of negotiation.

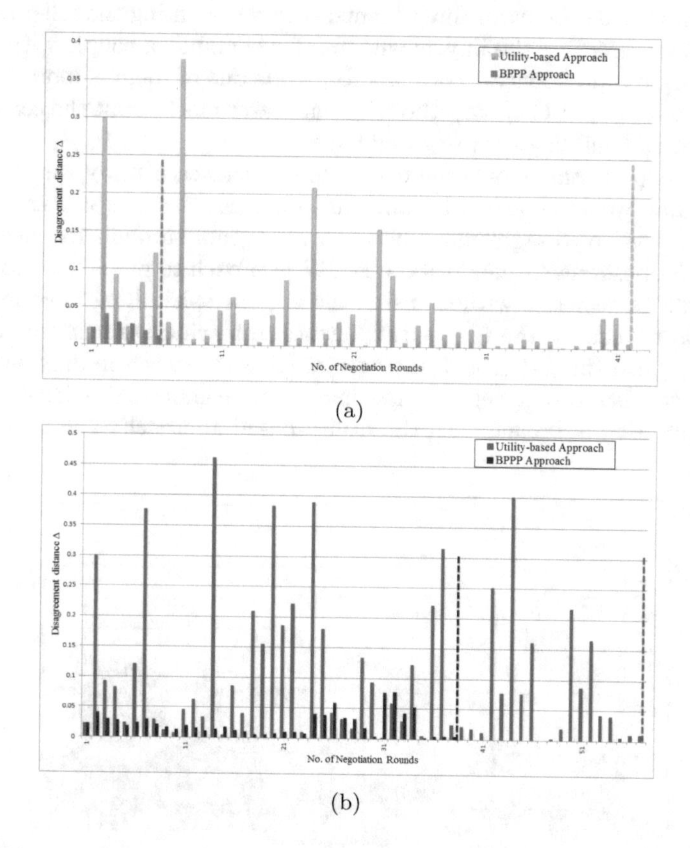

Fig. 10. Disagreement distances for a sequence of offered proposals in the Utility-based approach vs. the BPPP approach (a) With arguments (b) Without arguments. The dashed lines represent the end of negotiation.

concluded from both figures that the fluctuations in disagreement distances are smaller when we apply the BPPP technique. The dashed lines in the charts show where the negotiation ends for each setting.

3.4 Case Study B: King County House Sales

Although the first case study shows the efficiency of the proposed methodology, we introduced our second case study to ensure the performance, scalability, and applicability of the proposed methodology in other negotiation contexts with larger search spaces and higher dimensionalities. This dataset represents house sales in King County, US (Kaggle 2015). The data provide different attributes of the houses sold between May 2014 and May 2015. The database contains 21614 records, which let us test the performance and efficiency of our proposed techniques on a broad set of data. We also have a larger set of attributes including price, number of bedrooms, number of bathrooms, the house area (in sq ft), number of floors, house condition, year of built and landscape view. Here, the negotiation happens between three agents: One representing the seller agent and two others representing the buyers who are, for example, a couple with different perspectives and preferences. For instance, while one of them is more concerned about the price (e.g. PO agent), the other one cares more about the construction quality and the building area (e.g., QO agent).

Similar experiments have been conducted on this case study, and the results confirm what we have learned from our previous case study. Figure 11 represents the level of disagreement between all agents through the negotiation. Here, the proposer agent employs the BPPP approach to prepare a proposal for each round. As depicted, without using the argument-handling module, it takes around 500 rounds for the negotiating agents to reach an agreement. However, when we applied the AH model, the agreement was reached in only 34 rounds.

We have also run a set of experiments to compare the BP-based proposal preparation approach with the utility-based approach. The results of the

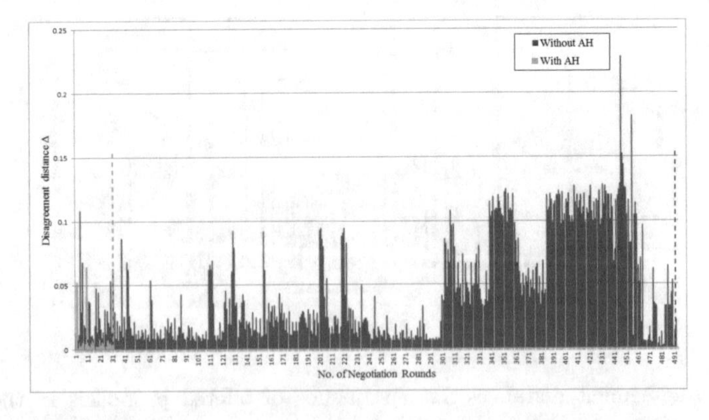

Fig. 11. Disagreement distances in negotiation over the second case study with and without AH model. The dashed lines represent the end of negotiation.

experiments, summarized in Fig. 12, confirms that the BPPP approach outperforms the utility-based approach when both of them benefit from the argument handling module. It takes 388 rounds for the negotiation with the utility-based approach to reach the agreement. However, the data shown in the chart is truncated to 300 rounds for improved legibility as the rest follows the same trend. The orange dashed line shows where the negotiation ends with the BPPP approach.

As Fig. 12 shows, there is a decreasing trend in the amount of normalized Δ when we use the utility-based approach. However, the result obtained from the BPPP approach does not follow any specific trend. In the utility-based approach, the proposer starts with his preferred attribute, which is the price attribute in this case study and ignores other attributes as well as other participant's preferences. Therefore, in the beginning of the negotiation, the level of disagreement is high. As the negotiation proceeds, the proposer improves its estimations about others' preferences through the arguments, and, therefore, the disagreement distance decreases. However, in the BPPP approach, the proposer starts as a neutral, giving all attributes equal importance. The only way he applies his own preferences is that he builds the unary cost tables based on his assumptions about the preferences of the other stakeholders. Then, the BPPP approach combined with the argument handling model allows him to adjust his assumptions in few rounds of negotiation.

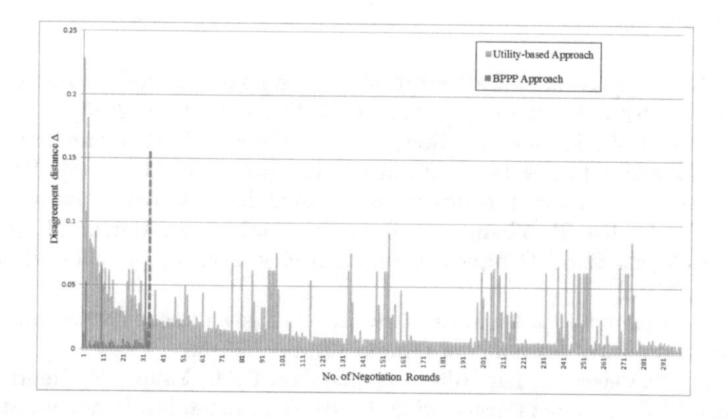

Fig. 12. Disagreement distances for a sequence of offered proposals in the Utility-based approach vs. the BPPP approach. The dashed lines represent the end of negotiation.

4 Conclusion and Future Work

This paper represents a novel negotiation strategy in which a belief-propagation based technique, as well as argument handling, are employed to improve the efficiency of the negotiation process. While the belief-propagation-based approach improves the proposal selection in each round of negotiation, the argument handling model provides the required information to feed this module. This information is gathered through agents communications about the past proposals and improve the proposer's learning rate to a great extent.To prove the ability of the developed methodology, a set of experiments has been conducted on two datasets, different in size and characteristics. The experiments on both environmental and non-environmental datasets confirm that argumentation-handling and the BP-based proposal preparation components facilitate and accelerate the negotiation process to a great extent. We have shown through experiments that the BP-based approach outperforms the utility-based approach regarding both the number of negotiation rounds and the fluctuations in disagreement distance measure.

In future, we will focus on more advanced models of stakeholders, where their decision-making process will be modeled using non-linear fuzzified functions. Also, the estimation process will be improved to adapt to the dynamic nature of the negotiations more appropriately. To build more complicated models for the stakeholders, more data ought to be gathered from real stakeholders; this will be done using a survey-based statistical technique such as the ones presented in Truong et al. (2015) and Cantillo et al. (2006).

Acknowledgements. This research project was partially supported by the funds available to the Schulich Chair in GeoSpatial Information Systems and Environmental Modeling held by Dr. Danielle Marceau at the University of Calgary.

References

Akhbari, M., Grigg, N.S.: A framework for an agent-based model to manage water resources conflicts. Water Resourc Manag **27**(11), 4039–4052 (2013)

Alfonso, B., Botti, V., Garrido, A., Giret, A.: A mas-based infrastructure for negotiation and its application to a water-right market. Inf Syst Front **16**(2), 183–199 (2014)

An, B., Lesser, V.: Yushu: a heuristic-based agent for automated negotiating competition. In. In: Ito, T., Zhang, M., Robu, V., Fatima, S., Matsuo, T. (eds.) New Trends in Agent-Based Complex Automated Negotiations, pp. 145–149. Springer, Heidelberg (2012)

Aragonès, E., Dellunde, P.: An automated model of government formation. In: Workshop on the Political Economy of Democracy, pp. 279–303 (2008)

Aydogan, R., Baarslag, T., Hindriks, K.V., Jonker, C.M., Yolum, P.: Heuristic-based approaches for CP-Nets in negotiation. In: Ito, T., Zhang, M., Robu, V., Matsuo, T. (eds.) Complex Automated Negotiations: Theories, Models, and Software Competitions, pp. 113–123. Springer, Heidelberg (2013)

Barbuceanu, M., Lo, W.K.: A multi-attribute utility theoretic negotiation architecture for electronic commerce. In: Proceedings of the Fourth International Conference on Autonomous Agents, pp. 239–246. ACM (2000)

Bäumer, C., Magedanz, T.: Grasshopper: a mobile agent platform for active telecommunication networks. In: Intelligent Agents for Telecommunication Applications (IATA 1999). Lecture Notes in Computer Science (Lecture Notes in Artificial Intelligence), vol. 1699, pp. 690–690 (1999)

Bousquet, F., Trébuil, G.: Introduction to companion modeling and multi-agent systems for integrated natural resource management in Asia. In: Companion Modeling and Multi-Agent Systems for Integrated Natural Resource Management in Asia, pp. 1–20 (2005)

Cantillo, V., Heydecker, B., de Dios, O.J.: A discrete choice model incorporating thresholds for perception in attribute values. Transp. Res. Part B Methodol. **40**(9), 807–825 (2006)

Cao, M., Luo, X., Luo, X.R., Dai, X.: Automated negotiation for e-commerce decision making: a goal deliberated agent architecture for multi-strategy selection. Decis. Support Syst. **73**, 1–14 (2015)

Chesnevar, C.I., Maguitman, A.G., Loui, R.P.: Logical models of argument. ACM Comput. Surv. **32**(4), 337–383 (2000)

Dworman, G., Kimbrough, S.O., Laing, J.D.: Bargaining by artificial agents in two coalition games: a study in genetic programming for electronic commerce. In: Proceedings of the 1st Annual Conference on Genetic Programming, pp. 54–62. MIT Press (1996)

El-Sisi, A.B., Mousa, H.M.: Argumentation based negotiation in multi-agent system. In: Proceedings of Seventh International Conference on Computer Engineering and Systems (ICCES), pp. 261–266. IEEE (2012)

Eshragh, F., Shahbazi, M., Far, B.: Automated dynamic negotiation over environmental issues. In: Proceedings of International Conference on Information Reuse and Integration, IRI 2017, San Diego, USA, pp. 92–99 (2017)

Faratin, P., Sierra, C., Jennings, N.R.: Using similarity criteria to make issue trade-offs in automated negotiations. Artif. Intell. **142**(2), 205–237 (2002)

Fink, A.: Supply chain coordination by means of automated negotiations between autonomous agents. Multiagent Based Supply Chain Manag. **28**, 351–372 (2006)

Freeman, R.E.: Strategic Management: A Stakeholder Approach. Cambridge University Press, Cambridge (2010)

Gilbert, M.A., Grasso, F., Groarke, L., Gurr, C., Gerlofs, J.M.: The Persuasion Machine, pp. 121–174. Springer, Dordrecht (2004)

Hadfi, R., Ito, T.: Low-complexity exploration in utility hypergraphs. J. Inf. Process. **23**(2), 176–184 (2015)

Hadfi, R., Ito, T.: On the complexity of utility hypergraphs. In: Fukuta, N., Ito, T., Zhang, M., Fujita, K., Robu, V. (eds.) Recent Advances in Agent-Based Complex Automated Negotiation, pp. 89–105. Springer, Cham (2016)

Jain, P., Dahiya, D.: An intelligent multi agent framework for e-commerce using case based reasoning and argumentation for negotiation. In: Dua, S., Gangopadhyay, A., Thulasiraman, P., Straccia, U., Shepherd, M., Stein, B. (eds.) Information Systems, Technology and Management, pp. 164–175. Springer, Berlin (2012)

Jennings, N.R., Corera, J.M., Laresgoiti, I.: Developing industrial multi-agent systems. In: Developing Industrial Multi-Agent Systems (ICMAS), pp. 423–430 (1995)

Jennings, N.R., Faratin, P., Johnson, M., Norman, T.J., O'Brien, P., Wiegand, M.: Agent-based business process management. Int. J. Coop. Inf. Syst. **5**(02n03), 105–130 (1996)

Jennings, N.R., Parsons, S., Norriega, P., Sierra, C.: On augumentation-based negotiation. In: Proceedings of the International Workshop on Multi-Agent Systems (IWMAS 1998), Boston, USA, pp. 1–7 (1998)

Jennings, N.R., Faratin, P., Lomuscio, A.R., Parsons, S., Wooldridge, M.J., Sierra, C.: Automated negotiation: prospects, methods and challenges. Group Decis. Negot. **10**(2), 199–215 (2001)

Kaggle: House Sales in King County, USA (2015). https://www.kaggle.com/harlfoxem/housesalesprediction. Accessed Aug 2017

Kakas, A., Moraitis, P.: Adaptive agent negotiation via argumentation. In: Proceedings of the Fifth International Joint Conference on Autonomous Agents and Multiagent Systems, pp. 384–391. ACM (2006)

Karunatillake, N.C., Jennings, N.R., Rahwan, I., McBurney, P.: Dialogue games that agents play within a society. Artif. Intell. **173**(9–10), 935–981 (2009)

Kraus, S., Lehmann, D.: Designing and building a negotiating automated agent. Comput. Intell. **11**(1), 132–171 (1995)

Kreps, D.M.: Game Theory and Economic Modelling. Oxford University Press, Oxford (1990)

Lomuscio, A.R., Wooldridge, M., Jennings, N.R.: A Classification Scheme for Negotiation in Electronic Commerce, pp. 19–33. Springer, Heidelberg (2001)

MacKenzie, A.B., DaSilva, L.A.: Game theory for wireless engineers. Synth. Lect. Commun. **1**(1), 1–86 (2006)

Marashi, E., Davis, J.P.: An argumentation-based method for managing complex issues in design of infrastructural systems. Reliabil. Eng. Syst. Saf. **91**(12), 1535–1545 (2006)

Okumura, M., Fujita, K., Ito, T.: An implementation of collective collaboration support system based on automated multi-agent negotiation. In: Ito, T., Zhang, M., Robu, V., Matsuo, T. (eds.) Complex Automated Negotiations: Theories, Models, and Software Competitions, pp. 125–1410. Springer, Heidelberg (2013)

Oprea, M.: Applications of multi-agent systems. In: Reis, R. (ed.) Information Technology, pp. 239–270. Springer, Boston (2004)

Ostrom, E., Gardner, R., Walker, J.: Rules, Games, and Common-Pool Resources. University of Michigan Press, Michigan (1994)

Parsons, S., Jennings, N.R.: Negotiation through argumentation a preliminary report. In: Proceedings of the 2nd International Conference on Multi Agent Systems, pp. 267–274 (1996)

Parsons, S., Sierra, C., Jennings, N.: Agents that reason and negotiate by arguing. J. Logic Comput. **8**(3), 261–292 (1998)

Pooyandeh, M., Marceau, D.J.: A spatial web/agent-based model to support stakeholders' negotiation regarding land development. J. Environ. Manag. **129**, 309–323 (2013)

Pooyandeh, M., Marceau, D.J.: Incorporating bayesian learning in agent-based simulation of stakeholders' negotiation. Comput. Environ. Urban Syst. **48**, 73–85 (2014)

Prakken, H., Vreeswijk, G.: Logics for Defeasible Argumentation, pp. 219–318. Springer, Dordrecht (2002)

Pruitt, D.G., Carnevale, P.J.: Negot. Soc. Confl. Thomson Brooks/Cole Publishing Co., Pacific Grove (1993)

Rahwan, I., Ramchurn, S.D., Jennings, N.R., Mcburney, P., Parsons, S., Sonenberg, L.: Argumentation-based negotiation. Knowl. Eng. Rev. **18**(4), 343–375 (2003)

Rahwan, I., Sonenberg, L., McBurney, P.: Bargaining and argument-based negotiation: some preliminary comparisons. In: International Workshop on Argumentation in Multi-Agent Systems, pp. 176–191. Springer (2004)

Ramchurn, S.D., Sierra, C., Godo, L., Jennings, N.R.: Negotiating using rewards. In: Proceedings of the Fifth International Joint Conference on Autonomous Agents and Multiagent Systems, pp. 400–407. ACM (2006)

Robu, V., Somefun, D., La Poutré, J.A.: Modeling complex multi-issue negotiations using utility graphs. In: Proceedings of the Fourth International Joint Conference on Autonomous Agents and Multiagent Systems, pp. 280–287. ACM (2005)

Shojaiemehr, B., Rafsanjani, M.K.: A supplier offer modification approach based on fuzzy systems for automated negotiation in e-commerce. Inf. Syst. Front. **20**(1), 143–160 (2018)

Sycara, K.P.: Negotiation planning: an AI approach. Eur. J. Oper. Res. **46**(2), 216–234 (1990)

Tamani, N., Mosse, P., Croitoru, M., Buche, P., Guillard, V., Guillaume, C., Gontard, N.: An argumentation system for eco-efficient packaging material selection. Comput. Electron. Agric. **113**, 174–192 (2015)

Thomopoulos, R., Croitoru, M., Tamani, N.: Decision support for agri-food chains: a reverse engineering argumentation-based approach. Ecol. Inform. **26**, 182–191 (2015). Information and Decision Support Systems for Agriculture and Environment

Truong, T.D., Wiktor, L., Boxall, P.C.: Modeling non-compensatory preferences in environmental valuation. Resour. Energy Econ. **39**, 89–107 (2015)

Van Kleef, G.A., De Dreu, C.K., Manstead, A.S.: Supplication and appeasement in conflict and negotiation: the interpersonal effects of disappointment, worry, guilt, and regret. J. Pers. Soc. Psychol. **91**(1), 124 (2006)

Von Neumann, J., Morgenstern, O.: Theory of Games and Economic Behavior. Princeton University Press, Princeton (1945)

Wang, M., Wang, H., Vogel, D., Kumar, K., Chiu, D.K.: Agent-based negotiation and decision making for dynamic supply chain formation. Engi. Appl. Artif. Intell. **22**(7), 1046–1055 (2009)

Wooldridge, M.: Intelligent agents: the key concepts. Multi-Agent Syst. Appl. **2322**, 3–43 (2001)

Zeng, D., Sycara, K.: Bayesian learning in negotiation. Int. J. Hum. Comput. Stud. **48**(1), 125–141 (1998)

Zhang, G., Jiang, G.R., Huang, T.Y.: Design of argumentation-based multi-agent negotiation system oriented to e-commerce. In: Proceedings of International Conference on Internet Technology and Applications, pp. 1–6. IEEE (2010)

Zhang, G., Sun, H., Jiang, G.: Adding argument into multi-agent negotiation in the context of e-commerce. In: IEEE Symposium on Robotics and Applications (ISRA), pp. 517–520. IEEE (2012)

SAIL: A Scalable Wind Turbine Fault Diagnosis Platform

A Case Study on Gearbox Fault Diagnosis

Maryam Bahojb Imani[1(✉)], Mehrdad Heydarzadeh[2], Swarup Chandra[1], Latifur Khan[1], and Mehrdad Nourani[2]

[1] Department of Computer Science, The University of Texas at Dallas, Richardson, USA
{maryam.imani,swarup.chandra,lkhan}@utdallas.edu

[2] Department of Electrical and Computer Engineering, The University of Texas at Dallas, Richardson, USA
{mehhey,nournai}@utdallas.edu

Abstract. Failure of a wind turbine is largely attributed to faults that occur in its gearbox. Maintenance of this machinery is very expensive, mainly due to large downtime and repair cost. While much attention has been given to detect faults in these mechanical devices, real-time fault diagnosis for streaming vibration data from turbine gearboxes is still an outstanding challenge. Moreover, monitoring gearboxes in a wind farm with thousands of wind turbines require massive computational power. In this paper, we propose a three-layer monitoring system: Sensor, Fog, and Cloud layers. Each layer provides a special functionality and runs part of the proposed data processing pipeline.

In the *Sensor* layer, vibration data is collected using accelerometers. Industrial single chip computers are best candidates for node computation. Since the majority of wind turbines are installed in harsh environments, sensor node computers should be embedded within wind turbines. Therefore, a robust computation platform is necessary for sensor nodes. In this layer, we propose a novel feature extraction method which is applied over a short window of vibration data. Using a time-series model assumption, our method estimates vibration power at high resolution and low cost. *Fog* layer provides Internet connectivity. Fog-server collects data from sensor nodes and sends them to the cloud. Since many wind farms are located in remote locations, providing network connectivity is challenging and expensive. Sometimes a wind farm is offshore and a satellite connection is the only solution. In this regard, we use a compressive sensing algorithm by deploying them on fog-servers to conserve communication bandwidth. *Cloud* layer performs most computations. In the online mode, after decompression, fault diagnosis is performed using trained classifier, while generating reports and logs. Whereas, in the offline mode, model training for classifier, parameters learning for feature extraction in sensor layer and dictionary learning for compression on fog servers and decompression are performed. The proposed architecture monitors the health of turbines in a scalable framework by leveraging the distributed computation techniques.

T. Bouabana-Tebibel et al. (Eds.): IEEE IRI 2017, AISC 838, pp. 96–118, 2019.
https://doi.org/10.1007/978-3-319-98056-0_5

Our empirical evaluation of vibration datasets obtained from real wind turbines demonstrates high scalability and performance of diagnosing gearbox failures, i.e., with an accuracy greater than 99%, for application in large wind farms.

Keywords: Big Data analytics · Gearbox
Industrial Internet of Things · Fault diagnosis · Signal processing
Vibration · Wind turbine

1 Introduction

1.1 Motivation

The wind energy industry has experienced major growth over the last decade. Based on Global Wind Energy Council statistics, the total number of wind turbines across the world has increased from 6,100 in 1996 to 3,695,789 in 2014 [26], showing an exponential growth in global demands for wind energy to produce electricity. Wind energy is freely available, does not pollute the environment, and needs less space for energy production.

Wind turbines are big, complex and expensive machines which are installed in harsh environments. They are vulnerable to different defects and faults. Manual condition monitoring is cumbersome and costly. A failure in these machines can disable them for several weeks and cost hundreds of thousands of dollars. These costs, typically due to maintenance and repair of failed components, adversely affect the price of produced energy. Developing automatic condition monitoring systems for these machines is vital for the wind industry. Based on a report by the National Renewable Energy Lab (NREL), gearbox failure causes the longest downtime, about half a month on average, and is the most costly fault in these machines [31]. Different factors like poor lubrication, bending fatigue, fretting corrosion, and mechanical stress can cause a defect in a wind turbine gearbox. An early fault detection can prevent catastrophic failures. Monitoring vibration signal picked up from gearboxes is an effective way of condition monitoring. However, by increasing the number of turbines, the number of sensors to monitor will dramatically increase. This typically leads to a large volume of data that needs to be processed. In fact, computation of these data cannot be achieved on site or locally. To address this challenge, we propose a cloud-based solution for the following reasons. First, wind turbines are mostly installed in harsh environments including off-shores and deserts. Providing a local computation platform for running a fault diagnosis algorithm requires a high level industrial specifications. Such a solution is expensive, specifically in typical cases where the fault diagnosis algorithm needs a powerful computation platform. Second, satellite links can provide network coverage in such remote areas; however, their services are still expensive, even with low bandwidth. Besides, wind turbines do not need continuous monitoring; regular maintenance with periods of a day or more is a generally an accepted practice. Finally, a local processing solution at every turbine is not appropriate to scale when monitoring wind farms with thousands of turbines. Therefore, monitoring each wind turbine with a standalone computer is not cost-effective.

1.2 Key Challenges

Even though increasing the volume of data provides more opportunities for accurate data analysis, it produces considerable computational challenges. Recently, the volume of data has been growing so fast in different areas that it can no longer fit into memory for efficient processing. To handle a huge amount of data, engineers have developed different scalable tools. New processing technologies such as Google's MapReduce and Apache Hadoop has emerged as a result of such efforts. With the exponential increase and availability of tremendous amounts of data, *Big Data* is becoming one of the most popular terms in today's automated world. From an industry point of view, Big Data is going to play an important role in achieving fault-free and cost-efficient data collection and analysis in real-time systems. Automated fault detection of wind turbines can be viewed as a Big Data problem. Sensor data from wind turbines can result in a continuous stream of data, useful for diagnosis. Unfortunately, traditional approaches [16] to address challenges of fault detection are inefficient in this setting. They typically require a high amount of memory and several hours to train an appropriate model and they cannot take advantage of a live stream of data. Using such traditional techniques may result in maintenance delays with increased costs.

From a signal processing perspective, there are still open research problems. Most of prior data-driven fault diagnosis algorithms are based on frequency domain analysis [8,14,29]. This is due to semi-periodic nature of vibration signal. Majority of prior works use the Fast Fourier Transform (FFT) to estimate the spectrum of signals which lead to a fast and an unbiased estimation of the spectrum. However, using FFT in feature extraction for an automatic fault diagnosis system raises a few challenges. First, FFT is computationally expensive and does not reduce the dimensionality of data. Moreover, the frequency resolution of FFT is limited to the number of the point used in its calculations. As a result, if the frequency signature of vibration for different classes are similar, FFT cannot capture them.

1.3 Contributions

Our contribution is twofold. First, we propose a scalable cloud-based computation platform denoted by *SAIL* (Scalable wind turbine fAult dIagnosis pLatform). Second, we propose a novel signal processing pipeline to diagnose the health state of a wind turbine. This algorithm runs on different computational elements in SAIL and is scalable. The key contributions of this paper are as follows:

- First, we propose a novel data-driven fault diagnosis algorithm. The proposed algorithm uses super-resolution spectral analysis for feature extraction from the vibration signal. We employ a time series model for vibration signal and

use the model parameters for feature extraction. These parameters represent the spectrum of vibration signals in a very compact way and require a reasonable computational cost. For classification, we apply the Random Forest algorithm which is in general suitable for parallel implementation.
- Second, we have developed a novel real-time framework for multi-source stream data to detect the turbine's faults using Apache Spark, with Kafka as a message broker [34]. In this framework, different sources can send their data stream through Kafka to a Spark cluster. Then, we apply the trained Random Forest model to analyze the data.
- Third, we address the class imbalance problem prevalent in applications such as fault diagnosis, unlike previous studies. Fault data is typically rare while fault-free data is available in abundance. Applying a traditional machine learning classifier over such biased data yields largely inaccurate model. We address this challenge by leveraging sampling techniques during data preprocessing.
- Fourth, we compare our fault detection technique by implementing it on traditional and Spark-based platforms. Our empirical results on real-world datasets show high detection accuracy with low latency. In addition, our proposed platform performance analysis indicates a significant reduction in computational time.

The paper proceeds with the prior works review in Sect. 2. This section also presents some descriptions about big data and Apache Spark as one of the big data analytics tools. Section 3 discusses about the platform architecture. Then, Sect. 4 shows the mathematical framework of feature extraction. Section 5 proposes the fault diagnosing algorithm, followed by a detailed description of how the proposed method works. Section 6 shows the experimental data and results on public-domain datasets. Finally, Sect. 7 presents conclusions with a few suggestions for future work.

2 Background

2.1 Industrial Applications

The two categories of fault diagnosis methods are model-based and data-driven. Model-based approaches use a mathematical model of the system [4]. Such a model is usually obtained by physical modeling. Having a model reduces the uncertainty regarding measurements, however, physical modeling is challenging. On the other hand, the only assumption that data-driven methods make is the availability of some training data from the system. Then, signal processing and machine learning are applied to predict the system's state of health [38]. Such approaches are easier to generalize compared to other approaches since they do not make any assumptions about the system.

The majority of prior gearbox fault diagnosis works are data-driven methods. However, since vibration signal is a non-stationary random signal, extracting a compact informative feature vector is a challenging signal processing task.

The key difference among prior works is in the feature extraction step. Authors in [19,28,41] proposed time-domain statistical features. Specifically, they showed that higher order statistics like kurtosis and skewness can be employed for fault diagnosis of the gearbox [18]. However, these features are very sensitive to noise and outliers. This limits their application to real-world industrial systems. The frequency domain features are proposed in [11,20,23,38]. Frequency domain methods are robust and can distinguish different classes of faults. However, for developing an automatic fault diagnosis system, it is necessary to reduce the dimensionality of signal representation. Fourier transform and wavelet transform do not necessarily reduce the dimension of a signal's spectrum. Authors in [3,36] post-processed the spectrum of the signal by Principal Component Analysis (PCA) for dimensionality reduction. However, PCA requires large computations at run-time and can corrupt the spectrum of the signal. So, presenting the spectrum of vibration signal in a compact way is required for developing robust fault diagnosis system. In a preliminary study, we developed a parametric spectral analysis for feature extraction using auto-regressive (AR) model [13]. Although, using AR model provides a method for compact representation, it requires a high model order for capturing complex signals. This issue can be addressed using a parametric models which has zeros. In this work, we used *autoregressive moving average (ARMA)* model which zeros and poles and can model vibration signal with smaller orders. Moreover, using ARMA model provides a high resolution in estimating spectrum of signal. This will be discussed in more details later.

The majority of research in this field is devoted to developing signal processing algorithms for fault diagnosis. However, this problem has another challenge for data analytics. Scaling a real-time fault diagnosis algorithm for a wind farm with thousands of turbines requires a massive computational power which is currently available on cloud servers. This dimension of research is related to the emerging field of *Industrial Internet of Things* (IIoT). From this perspective, authors in [5,27,30] applied the state-of-the-art Big Data analytics methods to the fault diagnosis problem in industrial systems in different areas.

2.2 Big Data Analytics

Many different Big Data analysis tools with different features have been developed so far, such as Hadoop [32] and Apache Spark [34]. In our work, we have selected Spark over other traditional distributed frameworks (e.g., Hadoop and MapReduce, etc.) since it is more efficient for stream data processing. Apache Spark is a fast and general purpose cluster computing engine for large-scale data analytics that was developed at the University of California, Berkeley AMP Lab [34]. Many different applications have been developed based on Spark clustering so far, political event coding [33], and geolocation extraction [12].

Another feature that Spark offers is supporting scalable, high-throughput, fault-tolerant, and real-time data stream processing. Spark streaming uses a "discretized stream" which is an incremental stream processing model. Data can be fed by many sources like Apache Kafka [15].

Machine Learning library (MLlib) is Spark's distributed and scalable machine learning library [22]. It improves the computational efficiency by using data-parallelism or model-parallelism techniques to store and process data or models. MLlib includes more than 50 common algorithms for classification, clustering, regression, collaborative filtering, and dimensionality reduction.

3 SAIL Architecture

3.1 Scalable Monitoring System

In the wind turbine industry, different sensors are being used to detect and predict various faults and malfunctions. As the number of turbine increases, the number of sensors, and consequently the amount of gathered data, grow as well. The traditional methods for data processing are time-consuming and inefficient for a huge amount of stream data. Therefore, Big Data analytics can play an important role in processing and mining these rapidly producing data in real-time.

3.2 Bottom-Up Architecture

We propose a bottom-up three-layer platform depicted in Fig. 1. At the bottom, the sensor-layer contains data collection nodes which are used to collect data from wind turbines. A node is attached to each turbine and provides sensory circuits and network connectivity. The next layer is the fog-layer which contains fog-servers. These servers collect data from all nodes in a wind farm.

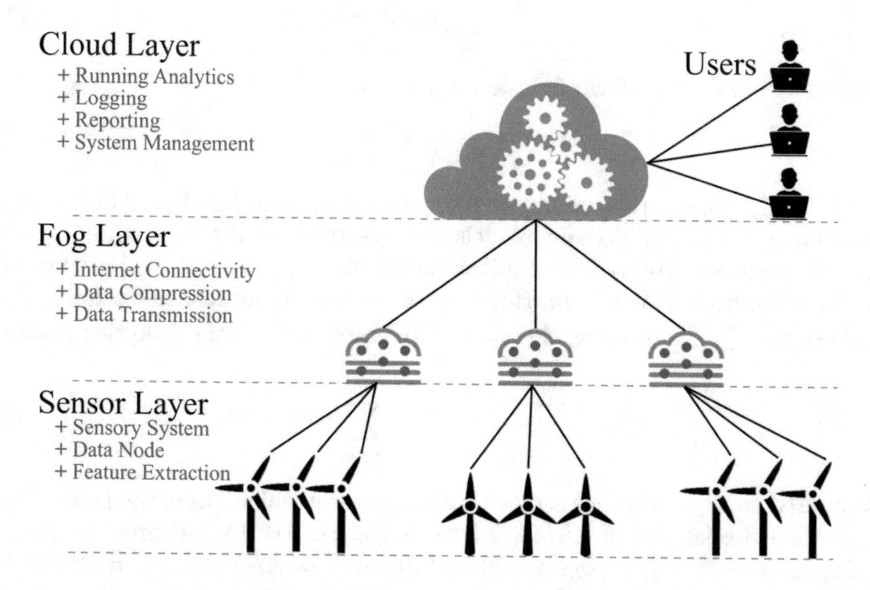

Fig. 1. The proposed three-layer platform (SAIL).

Fog-servers compress data to save the communication bandwidth. Then, they send the collected data to a cloud server via the Internet. In the cloud-layer, the analytic server runs the proposed fault diagnosis algorithm. After that, the results are logged and reported to different users. This architecture complies with the paradigm in the emerging field of Industrial Internet of Things (IIoT).

4 Mathematical Framework

4.1 ARMA Representation of Vibration Data

In this subsection, we introduce the mathematical framework underlying our proposed feature extraction method. We assume that vibration data is an ARMA time series. Then, we use parameters of this model for feature extraction.

Let $x[n]$ be a stationary time series. A time series model is a mathematical equation for predicting samples of time series. Assuming that each sample of $x[n]$ is a linear combination of past samples leads to the well-known *autoregressive (AR)* model. Mathematically, AR model is defined by $x[n] = \sum_{k=1}^{p} a_k x[n-k] + \epsilon[n]$ where a_k's are parameters of the model, p is model order and $\epsilon[n]$ is a zero mean white noise with variance of σ_ϵ^2. A discrete sinusoid function with frequency f_0, $x[n] = sin(2\pi f_0 n)$, can be written as a second order AR model without noise as $x[n] = a_1 x(n-1) + a_2 x(n-2)$ where $a_1 = 2cos(2\pi f_k)$ and $a_2 = 1$ with the initial condition of $x(-1) = -1$ and $x(-2) = 0$. By taking the Z transform, one can show that this model has a pair of poles at $f = \pm f_0$. Generally, a sum of p noise-free sinusoids, $x[n] = \sum_{i=1}^{p} sin(2\pi f_k)$, can be written as an AR model with order of $2p$:

$$x[n] = \sum_{k=1}^{2p} a_k x[n-k] \tag{1}$$

This model has a transform function in form of

$$H(z) = \frac{1}{1 + \sum_{k=1}^{2p} a_k z^{-k}} \tag{2}$$

which has $2p$ poles corresponding to frequency of sines at $f = \pm f_k$. This transfer function in frequency domain models the spectrum of the time series. If the sum of sines is corrupted by additive noise, like $y[n] = x[n] + \epsilon[n]$, the AR model assumption is not valid. However, one can write an AR model for $x[n] = y[n] - \epsilon[n] = \sum_{k=1}^{2p} a_k (y[n-k] - \epsilon[n-k]))$ which leads to the following model

$$y[n] = \sum_{k=1}^{2p} a_k y[n-k] + \sum_{k=0}^{2p} a_k \epsilon[n-k] \tag{3}$$

Compared to Eq. 1, this has a moving average term which filters the noise. This model is a special case of ARMA model. A general ARMA is defined by $y[n] = \sum_{k=1}^{p} a_k y[n-k] + \sum_{k=1}^{q} b_k \epsilon[n-k]$ and denoted as $ARMA(p,q)$. However, in Eq. 3 the moving average and the autoregressive parts are the same. By defining

$\boldsymbol{a} = [1, a_1, a_2, \cdots a_{2p}]^t$, $\boldsymbol{y} = [y[n], y[n-1]y[n-2]\cdots y[n-2p]]^t$ and $\boldsymbol{w} = [\epsilon[n], \epsilon[n-1], \epsilon[n-2], c\ldots, \epsilon[n-2p]]^t$ Eq. 3 can be written in matrix form as $\boldsymbol{y}^t\boldsymbol{a} = \boldsymbol{w}^t\boldsymbol{a}$. By pre-multiplying this equation by $\boldsymbol{y}$ and taking the expectation, one can get $E[\boldsymbol{y}\boldsymbol{y}^t]\boldsymbol{a} = E[\boldsymbol{y}\boldsymbol{w}^t]\boldsymbol{a} = E[(\boldsymbol{x}+\boldsymbol{w})\boldsymbol{w}^t\boldsymbol{a}] = E[\boldsymbol{w}\boldsymbol{w}^t]\boldsymbol{a}$. Intuitively, this is true since $\boldsymbol{a}$ and $\boldsymbol{x}$ are deterministic and $\boldsymbol{w}$ is zero mean and white. If $\boldsymbol{\Gamma}_{yy} = E[\boldsymbol{y}\boldsymbol{y}^t]$ denotes the auto-correlation matrix of $y[n]$, then, $\boldsymbol{\Gamma}_{yy}\boldsymbol{a} = \sigma_\epsilon^2\boldsymbol{a}$. Furthermore, it can be written as an eigenvalue problem as:

$$(\boldsymbol{\Gamma}_{yy} - \sigma_\epsilon^2\boldsymbol{I})\boldsymbol{a} = \boldsymbol{0} \tag{4}$$

where σ_ϵ^2 is eigenvalue and $\boldsymbol{a}$ is eigenvector. By solving Eq. 4, ARMA parameters and noise variance can be found. In practice, $\boldsymbol{\Gamma}_{yy}$ is not available and needs to be estimated which is discussed in the next sub-section.

4.2 Harmonic Decomposition

Consider a summation of p randomly phased sinusoid with white noise. Let A_i and $P_i = A_i^2/2$ denote the amplitude and power of the ith sinusoid in summation. Considering the white noise assumption, the auto-correlation function has a form of $\gamma(k) = \sigma_\epsilon^2\delta(k) + \sum_{i=1}^{p} P_i\, cos(2\pi f_i k)$. This auto-correlation can be written in a matrix form as:

$$\begin{bmatrix} cos(2\pi f_1) & cos(2\pi f_2) & \cdots & cos(2\pi f_p) \\ cos(4\pi f_1) & cos(4\pi f_2) & \cdots & cos(4\pi f_p) \\ \vdots & \vdots & \ddots & \vdots \\ cos(2p\pi f_1) & cos(2p\pi f_2) & \cdots & cos(2p\pi f_p) \end{bmatrix} \begin{bmatrix} P_1 \\ P_2 \\ \vdots \\ P_p \end{bmatrix} = \begin{bmatrix} \gamma_{yy}(1) \\ \gamma_{yy}(2) \\ \vdots \\ \gamma_{yy}(p) \end{bmatrix} \tag{5}$$

Having a set of samples of $y[n]$, there is maximum likelihood estimator for the auto-correlation function as $\hat{\gamma}_{yy}[k] = \sum_n y[n]y[n+k]$. If frequencies are known by calculating Eq. 5 and estimating $\hat{\gamma}_{yy}$, one can estimate sine powers using Eq. 5. The remaining problem is to determine the p frequencies which can be estimated by finding poles of Eq. 2 which in turn requires the knowledge of the eigenvector $\boldsymbol{a}$ in Eq. 4. For solving the eigenvalue problem in Eq. 4, the matrix $\boldsymbol{\Gamma}_{yy}$ can be formed using $\hat{\gamma}_{yy}$. For any arbitrary dimension of $\boldsymbol{\Gamma}_{yy}$ like $d \times d$, this matrix has d eigenvectors. However, the eigenvector equivalent to noise variance is needed which requires the knowledge of noise variance. One can show that the noise variance, σ_ϵ^2, is equivalent to the minimum eigenvalue of $\boldsymbol{\Gamma}_{yy}$ when $d \geq (2p+1)$ [35].

Finally, the power spectrum of the measured vibration signal ($y[n]$) can be estimated as following: First, estimate $\hat{\gamma}_{yy}[k]$ for $d = (2p+1)$ and form the matrix $\boldsymbol{\Gamma}_{yy}$. Then, eigenvalues of $\boldsymbol{\Gamma}_{yy}$ need to be found and the minimum eigenvalue is equivalent to σ_ϵ^2. The corresponding eigenvector holds the parameters of the model $ARMA(2p, 2p)$. Using these parameters, one can find poles of the transfer function in Eq. 2 which gives the frequencies. Having these frequencies and using Eq. 5, one can find amplitude of each frequency and thus the spectrum of the signal. This method is usually referred to as Pisarenko power spectrum estimation [35] which resolves frequencies with arbitrary accuracy not limited to $\frac{F_s}{2N}$. Due to this fact, model-based spectral analysis is sometimes referred to as *super resolution* method [35].

Generally speaking, the vibration data collected from a mechanical machine is not a stationary random process. However, the assumption of stationarity is valid over short windows of times. We verified this assumption in our work using a hypothesis test suggested by Kwiatkowski et al. [17]. The details of this test is beyond the scope of this paper. Briefly, we propose a grid search mechanism over different values of window length, we choose the smallest length for which the hypothesis test does not fail.

4.3 Model Order Selection

The above-mentioned method estimates the spectrum lines using an ARMA model for vibration data. However, in developing this model it is assumed that the model order or number of sinusoids is known. In practice, the exact number of sinusoids (p) is not known. However, from the physical modeling perspective, one can estimate p by statistical methods.

For a fitted model, one can calculate $\hat{y}[n] = \sum_{k=1}^{p} A_k sin(2\pi f_k n)$ and form the residual signal as $e[n] = y[n] - \hat{y}[n]$, which is equivalent to our estimation for noise $\epsilon[n]$. However, since $\epsilon[n]$ is assumed white zero mean Gaussian, one can calculate the likelihood of the model as $\mathcal{L} = \prod_i [(2\pi\sigma_\epsilon^2)^{-1/2} exp(e[n]/\sigma_\epsilon)^2]$ with the assumption of independent and identical distribution. In a good estimation, each sample of residual, $e[n]$ is close to zero and its probability becomes bigger and as a result, the probability of all samples becomes large.

A higher p leads to a higher likelihood and lower error. However, a too high p is too sensitive to noise. For balancing this trade-off, the Akaike information criteria (AIC) [35] is used. AIC puts a penalty term on the likelihood ($\mathcal{L}$):

$$AIC = 2p - 2ln(\mathcal{L}) \tag{6}$$

which penalizes a $2p$ term for number of parameters in model. Given a set of candidate model from data, the model with minimum AIC is preferred. Here, we sweep an interval for p around our initial guess based physical modeling and a model order with minimum AIC will be chosen.

4.4 Dictionary Learning and Compression

A bottleneck in proposing a cloud based solution for monitoring wind-farms is the connectivity bandwidth. Majority of wind-turbines are installed offshore where providing the network connectivity is challenging and expensive. The sensor node extracts ARMA coefficients over rolling window of vibration data. The stream of ARMA coefficients is fed to fog server. Here, the fog server compresses this stream and sends them to the cloud. In this section, the proposed compression algorithm is reviewed.

The basic concept in compression is assigning shorter codes to symbols with higher frequencies such as the well-known Huffman coding algorithm. It is very useful to transform a signal to another domain which leads to a more sparse representation. It has been proven that some transforms like Fourier or Wavelet provide a sparse representation of broad group a time series [21]. Such transforms are exactly or approximately invertible, which provides a mechanism for reconstructing the original time series after decompression. Applying a variable length coding algorithm to a more sparse representation provides a higher compression gain which is desirable. This gain is usually traded off by computation cost of transformation.

Generally, the transformation is done by decomposing the signal as a linear combination of basis functions (*atoms*). In the Fourier transform, these basis functions are complex exponential. In wavelet transform, different wavelet basis functions are proposed. Usually, the set of basis functions, so called *dictionary*, is an orthogonal set which preserves some sort of completeness. Orthogonal basis functions are useful since the inner product operation can be used for finding the decomposition coefficients. A set of n real orthogonal basis functions form a complete basis in $\mathbf{R}^n$, if any arbitrary signal with length n like $\boldsymbol{x} \in \mathbf{R}^n$ can be uniquely represented using the elements of the set. In many traditional compression applications, these basis functions are chosen from existing dictionaries like Fourier or wavelet dictionaries. However, it is possible to design a dictionary for a class of signal to get a better compression rate. In this sub-section, we briefly review the concept of sparse decomposition and the proposed compression algorithm foundation. Here, the term signal refers to the stream of ARMA coefficients extracted in sensor node.

4.5 Sparse Decomposition

Let $\boldsymbol{y} \in \mathbf{R}^n$ denotes a signal with length of n. For a given dictionary of signal atoms like $\boldsymbol{D}$, this signal can be represented as $\boldsymbol{y} = \boldsymbol{D}\boldsymbol{x}$, where $\boldsymbol{x} \in \mathbf{R}^K$ holds the signal representation coefficients in a new domain where K is the number of atoms. If $\boldsymbol{y} = \boldsymbol{D}\boldsymbol{x}$ is satisfied the representation is called exact and the compression is loss-less. In a lossy compression scheme, the representation can be an approximate $\boldsymbol{y} \approx \boldsymbol{D}\boldsymbol{x}$, and the error is bounded by satisfying a condition like $\|\boldsymbol{y} - \boldsymbol{D}\boldsymbol{x}\|_p \leq \epsilon \ \|\boldsymbol{x}\|_p$, where $\|\boldsymbol{x}\|_p = (|x_1|^p + \cdots |x_n|^p)^{1/p}$ is the l_p norm. In approximation applications, the l_2 norm or Euclidean norm is usually used. Each column of the dictionary matrix $\boldsymbol{D} \in \mathbf{R}^{n \times K}$ is a basis function (atom) where $\boldsymbol{d}_j \in \mathbf{R}^n$ and $\|\boldsymbol{d}_j\|_2 = 1$ for $j = 1, \cdots, K$. If $n < K$ and $\boldsymbol{D}$ is a full rank matrix, the dictionary is called over-complete and there is an infinite number of solutions for $\boldsymbol{x}$. One may choose $\boldsymbol{x}$ such that it results in a sparse solution, leading to a simpler representation. The sparsity can be quantized using l_0 norm which is equivalent to number of nonzero elements of a vector. So, the sparse decomposition can be found by solving the following optimization:

$$\boldsymbol{x}^* = \arg\min_{\boldsymbol{x}} \|\boldsymbol{x}\|_0 \qquad \textit{subject to } \boldsymbol{y} = \boldsymbol{D}\boldsymbol{x} \tag{7}$$

where $\|\boldsymbol{x}\|_0$ is the number of nonzero elements in $\boldsymbol{x}$. In approximate representation, the constraint in Eq. 7 will be replaced by $\|\boldsymbol{y} - \boldsymbol{D}\boldsymbol{x}\|_2 \leq \epsilon$.

If the dictionary is orthogonal, the decomposition coefficients can easily be obtained by inner product operator. However, for sparse decomposition of a signal on an over-complete dictionary requires solving Eq. 7. An optimal algorithm for solving the problem in such a scenario is *Orthogonal matching pursuit* (OMP) [6]. OMP has two steps in each iteration: (i) sweep (ii) update. These two steps are applied in a loop to minimize the residual of signal which is the difference of the signal and its sparse approximation. In the first iteration, in the sweep step, the inner product of the signal and all dictionary atoms are calculated and the atom with largest absolute value of inner product is selected. In the update step the residual of signal as the difference between ***argument*** y and $\boldsymbol{D}\boldsymbol{x}$ will be calculated and is projected to the space orthogonal to span of all selected atoms.

4.6 Dictionary Learning Using K-SVD

In this sub-section the problem of dictionary learning is discussed which is an unsupervised problem. The proposed monitoring system uses K-SVD algorithm [2]. This algorithm is run on cloud to extract the compression dictionary, D. Then, the learned dictionary is sent to fogs for running the compression task.

Basically, K-SVD is a generalization of the well-known *k-means clustering* algorithm. It uses singular value decomposition (SVD) method in linear algebra to reduce the error of learning atoms of a dictionary. Suppose a set of N training signals like $\{\boldsymbol{y}_1, \cdots, \boldsymbol{y}_N\}$ is given. This set can be represented as a matrix form of $\boldsymbol{Y} = [\boldsymbol{y}_1, \cdots, \boldsymbol{y}_N]$ which is an $n \times N$ matrix. Each $\boldsymbol{y}_i$ in this set has a sparse representation like $\boldsymbol{x}_i$. Let $\boldsymbol{X} = [\boldsymbol{x}_1, \cdots, \boldsymbol{x}_N]$ denote the corresponding sparse representation of given training set. In Eq. 7 the sparse representation of a single vector is obtained by an optimization. Here, the dictionary needs to be chosen such that this representation is as sparse as possible for all vectors. So, the dictionary can be found by minimizing the sum of objective function in Eq. 7 for all vectors as $\arg\min_{\boldsymbol{D},\boldsymbol{X}} \sum_i \|\boldsymbol{x}_i\|_0$ subject to $\|\boldsymbol{y}_i - \boldsymbol{D}\boldsymbol{x}_i\|_2 \leq \epsilon$ for $i = 1, \cdots, N$. Using the matrix form, this cost can be equivalently written as:

$$< \boldsymbol{D}^*, \boldsymbol{X}^* >= \arg\min_{\boldsymbol{D},\boldsymbol{X}} \|\boldsymbol{Y} - \boldsymbol{D}\boldsymbol{X}\|_F \qquad \textit{subject to} \; \forall i, \|\boldsymbol{x}\|_0 \leq T_0 \tag{8}$$

where $\|\cdot\|_F$ is the Frobenius norm which is simply sum of square of all elements. Minimizing the objective function in Eq. 8 minimizes both the error for choosing the sparse representation of a vector and choice of the dictionary atoms.

K-SVD is an iterative algorithm which has two steps in each iteration: (i) Sparse coding step and (ii) codebook update step. In the sparse coding step, the approximate sparse representation of all training vectors, $\boldsymbol{Y}$, are obtained using a pursuit algorithm like OMP by solving the optimization in Eq. 7 to obtain $\boldsymbol{X}$. Next, in the codebook update stage, a single atom in $\boldsymbol{D}$, let say $\boldsymbol{d}_k$, will be updated. For this purpose, all vectors in $\boldsymbol{X}$ which used $\boldsymbol{d}_k$ are grouped together. Then, the error of choosing $\boldsymbol{d}_k$ is minimized. For this purpose the SVD is used to find a rank 1 estimation of this error. In the next iteration, another atom in the dictionary will be updated. The details of K-SVD is beyond the scope of this paper and reader can refer to [2] for more details.

5 Fault Diagnosis Framework

In this section, the proposed scalable architecture is discussed. Figure 2 shows the multi-layer structure of SAIL. This figure shows where each piece of algorithm runs. Training algorithms run offline, they are depicted as gray boxes. Other modules work and process data online. Development of a mechanic fault can take minutes or even hours so the whole system can generate early alerts in an online manner. There are three layers in the proposed system: sensor-layer, fog-layer, and cloud-layer. Furthermore, each layer contains several components which are discussed in this section.

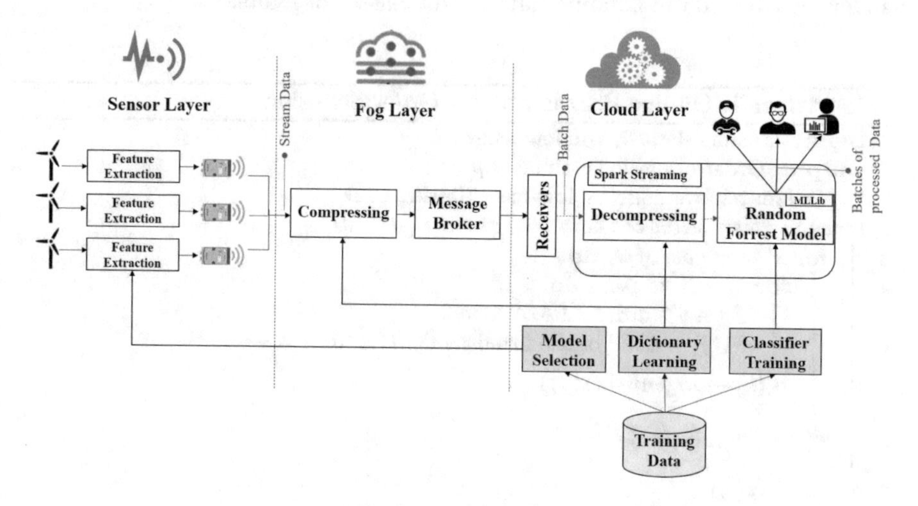

Fig. 2. The proposed fault diagnosing framework

5.1 Sensor Layer

In this layer, each turbine in a farm has a node which provides sensory circuits and local network connectivity. In each node, vibration data is collected using accelerometers. After extracting the features, it broadcasts the data stream to the fog layer.

5.1.1 Feature Extraction

We applied the model-based spectral analysis method explained in Sect. 4 for feature extraction. We considered an ARMA model for the vibration signal and use the parameters of the model as features for diagnosing the health state of the system. These features capture the spectrum of the signal and simultaneously reduces its dimensionality. The proposed system uses a p order on-line ARMA estimator to a sliding window of vibration signal from the device to extract coefficients. Then, our features are the vector of all model coefficients $\theta = [a_1, a_2, \cdots, a_p]^t$. Vector $\boldsymbol{\theta}$ is fed to a trained random forest classifier to diagnose the status of the device.

Here, the order of online estimators is unknown and needs to be estimated. As discussed in Subsect. 4.3, AIC criteria can be used for order selection. However, we do not deal with just one random process because each class of fault produces a different random process as vibration signal. Consequently, each class has its own order. The order of the online estimator should be chosen such that the fitted model captures the spectral properties of the vibration signal for each fault. So, we set this parameter as the largest estimated p in training data for all classes of vibration data. The pseudo code illustrated in Algorithm 1 shows the algorithm used for order selection which finds the maximum order for each window of vibration in training data for all classes of faults.

Algorithm 1. On-line PSD Estimator Order Selection

Input: Training data $\mathbb{D}$, window length L
Output: Order of on-line estimator p

1 **Initialize**: remove noise by low-pass filtering of $\mathbb{D}$
2 **foreach** *class, select the data in that class* ($\mathbb{D}_i$) **do**
3   **for** k^{th} *window of* $\mathbb{D}_i$ **do**
4   **for** $j \leftarrow 1$ *to* p_{max} **do**
5   Fit a jth order ARMA model
6             $F_k(j) \leftarrow$ AIC of the fitted model (Eq. 6)
7         $p_i(k) \leftarrow \underset{j}{argmin}\{F_k(j)\}$
8     $p(i) \leftarrow \underset{k}{max}\{p_i(k)\}$
9 $p \leftarrow \underset{i}{max}\{p(i)\}$
10 **return** p *as the order of on-line estimator*

5.2 Fog Layer

This layer compresses the collected data from the wind turbine farm to save communication bandwidth. As we discussed dictionary learning and compression in Subsect. 4.4, the dictionary learning is offline since it uses the training data. However, the data compression is online for signals coming from wind turbines. Then, this layer sends the collected compressed-data to the next layer, i.e. cloud layer, using Kafka as a message broker.

5.2.1 Message Broker Module

Various message brokers are available to integrate with Spark. We have applied Kafka 3.3.4 because of its stability and compatibility with Apache Spark. Kafka is a real-time publish-subscribe messaging system developed by LinkedIn [15]. Kafka publisher-clients write messages in topic categories, each topic category is divided into several partitions, and messages within a partition are totally ordered. Kafka subscriber clients read messages on a topic. It provides reliable message delivery with the correct order. Our framework continuously monitors each incoming datum from different sources. We read these streaming data and transport it through a message broker. It enables us to feed these large volumes of raw data to the Spark streaming module (Fig. 2).

5.3 Cloud Layer

Initially, we perform model and parameter selection for feature extraction (Subsect. 5.1.1) and train a diagnosis model offline by using raw training data. Moreover, dictionary learning for compressing and decompressing is done in offline mode. Then, the prepared model will be used in the online fault diagnosing framework, which is presented in Fig. 2.

Algorithm 2. Random Forest Classifier.

Input: Training data ($\mathbb{D}$), number of trees (B), Minimum node size (N_m)
Output: A random forest classifier
1 **for** $i \leftarrow 1$ *to* B **do**
2     $\mathbb{Z} \leftarrow$ a random subset of $\mathbb{D}$
3   Grow a random tree (T_b) using $\mathbb{Z}$ as:
4   **for** $j \leftarrow 1$ *to* N_m **do**
5   Select m variable at random from p variable
6   Pick the best variable/ split-point among the m
7   Split the node into two daughter
8 **return** *The ensemble of trees* $\{T_b\}_1^B$

In this layer, the Spark Streaming module receives a stream of data and convert them to batches of data. After decompressing the data, we use the classifier to diagnose whether there is faulty data or not (Subsect. 5.3.1).

To evaluate the model's performance, we show the experimental results with offline data in Sect. 6. In this section, we will introduce our feature extraction method and machine learning algorithm which are used in this study.

5.3.1 Random Forest Classifier

Decision trees are simple classifiers which make an output label by comparing different features with threshold values in a tree structure. Although trees are unbiased classifiers, their variance is high. One way of decreasing their variance and thus increasing the classification accuracy is by *bagging*, which combines the output of several trees [10]. An ensemble of several trees is usually referred to as a *forest.* In a forest, each tree votes for the label of an input and the final decision is made using the majority voting technique. Although each tree is a simple classifier, the forest can learn difficult classification tasks by bagging trees together. One advantage of a forest is its suitable structure for real-time implementation since each can be evaluated in parallel and the final decision can be made by combining all results together. *Random Forest* (RF) is an effective way of training bagged trees where tree parameters are chosen randomly [10]. In a random forest, each tree is trained using a random subsample of training data. Algorithm 2 shows the random forest algorithm. In this study, we used RF in MLlib which supports the Random Forest's parallelization.

6 Experimental Results

We present two categories of experiments to evaluate SAIL platform and the proposed fault diagnosing algorithm in this section. First, we measure the accuracy of the proposed fault diagnosing method by applying it to two well-known vibration benchmark datasets, i.e. NREL and CWRU datasets. Using benchmark datasets provide an opportunity to compare our algorithm with prior works regardless of our scalable implementation. Next, we evaluate the scalability and run-time of the proposed platform. In this experiment, we created multiple sensor node instances which feed duplicate of these datasets to the SAIL.

6.1 Algorithm Accuracy Test

6.1.1 NREL Dataset

The vibration data used in this research is provided by the National Renewable Energy Lab (NREL) [24]. The test turbine is a 750 kW three-bladed upwind turbine with stall control. Figure 3 shows the turbine.

For testing the proposed algorithm and platform, we used the wind turbine gearbox data provided by NREL. This dataset is collected from a damaged gearbox. The complete bed plate and drive-train of a turbine were installed at dynamo-meter test facility (DTF) of NREL [24]. The bed plate is fixed to the floor without the rotor, yaw bearing, or hub. The gearbox is shown in Fig. 3 and has three-stages: low-speed stage (LSST), intermediate-speed stage (ISST) and high-speed stage (HSST), as shown in Fig. 4. LSST is connected to the rotor and HSST is connected to the generator. In a wind turbine, the LSST shaft is connected to the rotor and the HSST shaft is connected to the generator.

For measuring the vibration data, a set of accelerometer sensors are mounted to sides of the gearbox housing. On the side, the speed of HSST shaft is recorded. In this study, we only use the vibration data. The vibration data of each sensor is sampled with a PXI - 4472B data acquisition board from National Instruments at the rate of 40 KHz per channel. In this paper, $V_i[n]$ for $i = 1, \cdots, 8$ denotes

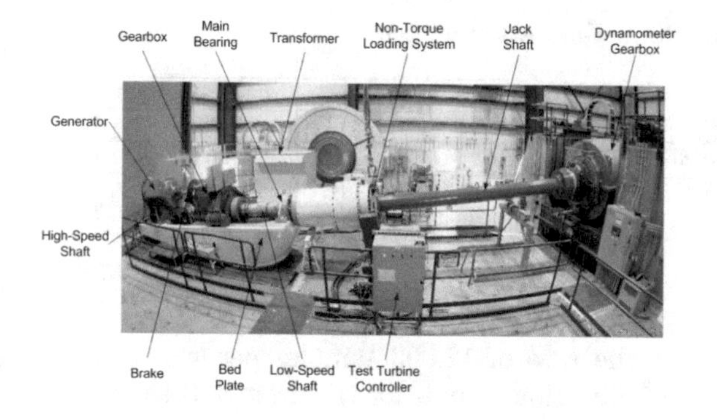

Fig. 3. The test turbine (image from [24])

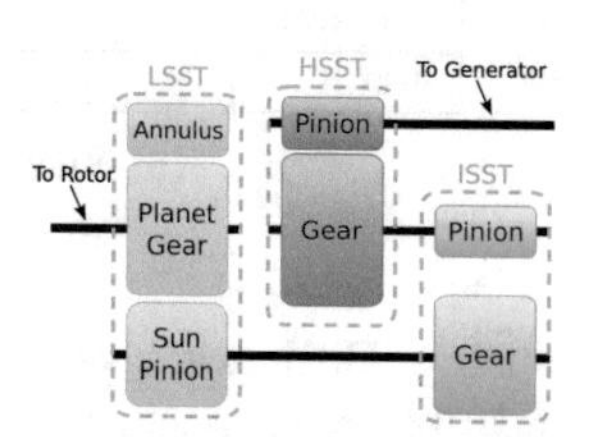

Fig. 4. The gearbox block diagram

the samples of the ith accelerometer for eight available sensors. The dataset is publicly accessible on request through [24].

6.1.2 CWRU Dataset

We apply SAIL to the vibration data provided by the Case Western Reserve University (CWRU) bearing data center [1] as a benchmark. CWRU dataset provides a number of faults in bearing of gearboxes. Using these benchmarks provide a fair and standard framework for comparison of our algorithm with different algorithms. The experimental setup, depicted in Fig. 1a, consists of a 2 HP motor, a torque transducer in the center, and a dynamometer as a load. Bearings of motor shaft are studied in this benchmark. Figure 5 shows the setup's block diagram.

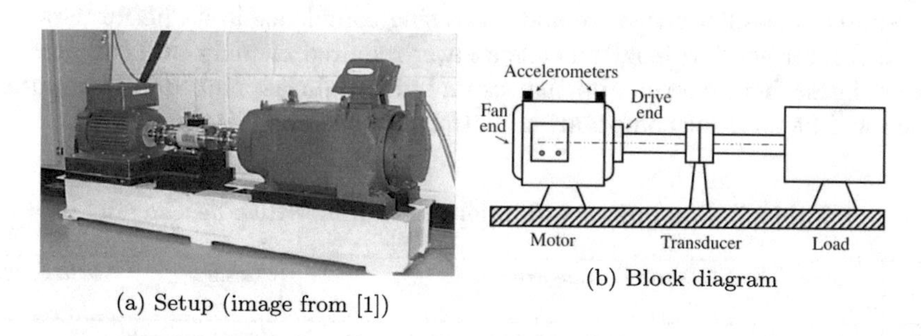

(a) Setup (image from [1])

(b) Block diagram

Fig. 5. Experimental setup of CWRU bearing data [1].

Single point faults were introduced to test bearings using an electro-discharge machine with diameters of 7, 14 and 21 mils on different parts of bearing including the inner race, the outer race, and the ball. Vibration data was collected using two single axial accelerometers with magnetic bases. One accelerometer was mounted on the drive-end and the other on the fan-end of the motor's

Table 1. Datasets statistics

Data Status	# Instances
Normal	93501
Fault	103890

(a) NREL

Data Status	# Instances
Normal	8844
Ball Fault	15177
Inner Race Fault	15179
Outer Race Fault	31019

(b) CWRU

housing. Both were sampled at the rate of 12,000 samples per second in the constant shaft speed. Samples of vibration signals for the drive and the fan end accelerometers, respectively.

The experiment was repeated for different fault intensities (7, 14 and 21 mils) and locations (ball, races and drive/fan ends). Moreover, data were collected for four constant load conditions (0, 1, 2 and 3 HP) for each class of faults. This dataset has four classes of vibration data: fault-free, inner-race (IR) fault, outer-race (OR) fault and ball fault.

The overall statistics of these datasets, i.e. NREL and CWRU are presented in Table 1.

6.1.3 Accuracy and Comparison

We apply a 10-fold cross validation to report experimental results. The accuracy of SAIL is compared with prior work on NREL and CWRU dataset as reported in the literature in Tables 2 and 3, respectively. These tables clearly demonstrate that SAIL outperforms the existing methods in performance. The proposed method can capture a small difference of vibration spectrum with high accuracy. Moreover, SAIL processes the vibration data in shorter windows. We propose a shorter window length and a compact feature vector which results in a simpler classifier structure and increasing confidence in its performance. In fact, a shorter window length means a lower required memory and computation power. These factors form a bottleneck on industrial-class embedded computers, which will be used for computation platform at the node level.

Table 2. Comparison with prior works using NREL benchmark

Ref.	Window size	# of Features	Feature extraction	Classifier	Accuracy (%)
[41]	40,000	N/A	Time and frequency clustering	Neural network	98.46
[40]	8,000	375	Side band power factor	SVM	90
[13]	512	80	Reflection coefficients of AR model	RF	98.93
SAIL	512	48	**Model-based spectral analysis**	RF	**99.53**

Table 3. Comparing various fault diagnosis methods using CWRU benchmark

Ref.	Window size	# of Features	Type of features	Classifier	Accuracy (%)
[9]	12,000	45	Global spectrum	Linear classifier	95
[18]	4,096	24	Statistical & frequency domain	Multiple ANFIS	91.33
[3]	4104	4104	Vibration Spectrum Imaging of FFT	ANN	96.90
[28]	N/A	80	Two morphological operators on both FFT and Wavelet	Fuzzy classifier	N/A
[11]	4,096	11	Selected from 98 statistical & Wavelet packet features	Ensemble SVM	90
[39]	N/A	10	Selected from a pool of 57 features in different domains	Fuzzy ARTMAP	89.63
[19]	35,750	20	Nearest-farthest distance preserving projection of statistical features	N/A	99.75
SAIL	**1024**	**21**	**Model-based spectral analysis**	**RF**	**99.15**

6.1.4 Data Imbalance Problem

These datasets contain almost the same number of fault instances as normal instances. However, in the real world, the turbine's fault and normal data are not balanced due to the presence of few fault cases compared to a large number of normal instances. In other words, in wind turbines, the fault instances are always rare compared with occurrences of normal instances when the turbine is fully operational. As a result, the dataset becomes highly imbalanced for directly applying classification techniques. In such cases, standard classifiers tend to be overwhelmed by the large classes and ignore the small ones. For a typical fault (e.g., gearbox faults, rotor imbalance, blade angle asymmetry) the ratio of normal to fault instances can be as high as 1000:1 [37]. Different solutions such as oversampling (O) and undersampling (U) approaches have been applied to the class-imbalance (Imb) problem. In oversampling, the instances of the minority class are randomly duplicated until a fully-balanced dataset is realized. In undersampling, instances of the majority class are randomly discarded from the dataset until a full balance is reached [7].

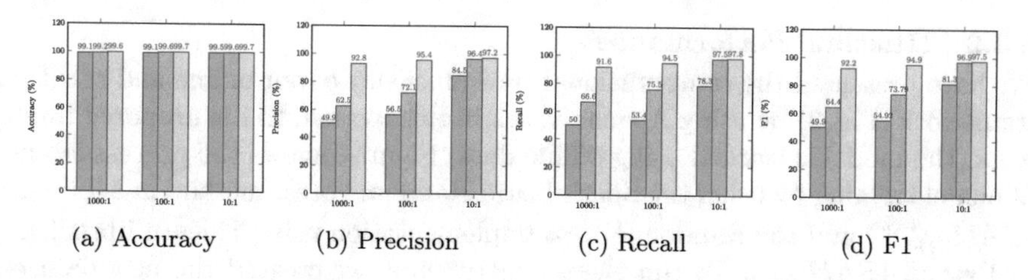

(a) Accuracy (b) Precision (c) Recall (d) F1

Fig. 6. Performance of $SAIL_{Imb+RF}$ with different ratios of normal to fault data instances, compared to $SAIL_{U+RF}$ and $SAIL_{O+RF}$ on NREL dataset

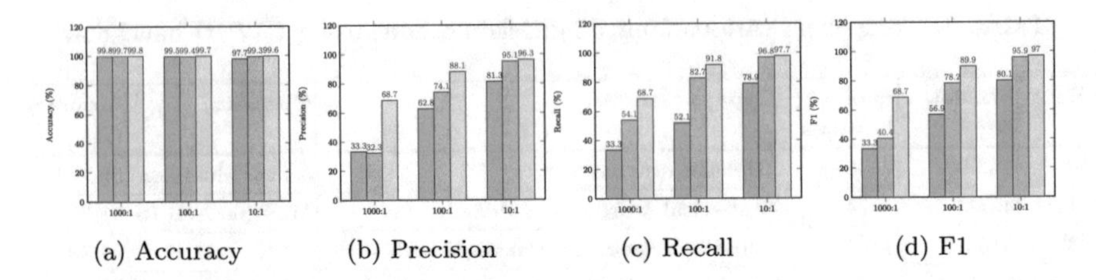

(a) Accuracy (b) Precision (c) Recall (d) F1

Fig. 7. Performance of $SAIL_{Imb+RF}$ with different ratios of normal to fault data instances, compared to $SAIL_{U+RF}$ and $SAIL_{O+RF}$ on CWRU dataset

Figures 6 and 7 show the experimental results of SAIL for different imbalance ratios on NREL and CWRU dataset, respectively. In $SAIL_{Imb+RF}$, without using imbalance solution technique, the RF classifier overfits with the normal class and ignores the fault class(s). As a result, $SAIL_{Imb+RF}$ does not perform well. Therefore, we address this problem using $SAIL_{O+RF}$. This works better than $SAIL_{U+RF}$ since, in the latter technique, the RF classifier could not learn a good discriminative model with fewer instances of the normal class.

6.2 Platform Scalability Test

6.2.1 Cluster Setup

Our VMware cluster consists of 5 VMware ESXi [Palo Alto, CA] (VMware hypervisor server) 5.5 systems. Each of the systems has Intel(R) Xeon(R) CPU E5-2695 v2 2.40 GHz processor, 64 GB DDR3 RAM, 4 TB hard disk, and dual NIC card. Each processor has 2 sockets and every socket has 12 cores. So there are 24 logical processors in total. All of the ESXi systems contain 3 virtual machines. Each of the virtual machines is configured with 8 vCPU, 16 GB DDR3 RAM and 1 TB Hard disk. As all the VM's are sharing the resources, performance may vary in runtime. We have installed Linux Centos v6.5 64 bit OS in each of the VM along with the JDK/JRE v1.8. We have installed Apache Spark version 2.1.0. We have also installed Apache Hadoop NextGen MapReduce (YARN) with version Hadoop v2.6.0 and formed a cluster.

6.2.2 Runtime Performance

We have presented the computational time efficiency of our offline and real-time framework (Fig. 2) by using Apache Spark in this section. In the proposed framework, the model is learned using offline data. Figure 8 shows the process latency of model training by using the Spark-based Random Forest method in MLlib [22] ($SAIL_{RF}^{Spark}$) and the Random Forest implementation using Sklearn library [25] in Python ($SAIL_{RF}$). To run these experiments, we created the new datasets by duplicating the original datasets for α times, where $\alpha \in \{5, 10 \ldots 30\}$, to show training latency by increasing the volume of the data. We have measured the process latency of the model after a certain number of training instances

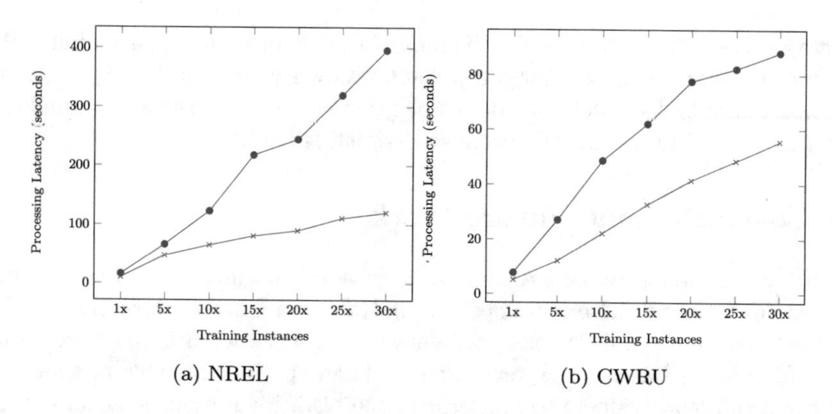

Fig. 8. Comparing processing latency during training Random Forest model with Mllib in Spark ($SAIL_{RF}^{Spark}$) —×— and Sklearn library ($SAIL_{RF}$) —●— using NREL and CWRU dataset.

have been processed and plotted them. Figure 8 shows that Spark-based approach performs significantly better than traditional methods when the number of training instances is increased.

6.2.3 Parallelism Test

We have utilized a message broker using Kafka, which is highly available and scalable. It helps us to add more sources to behave as turbine sensors to collect more test data. We vary the number of source instances while we collect data. In fact, these source instances simulate wind turbines to generate data. All of these sources work in parallel and independent of each other. So when the number of sources increases, more data will be produced. Figure 9 shows the average process latency of fault diagnosis of online data. In this figure, we see that the average process latency of fault diagnosis does not increase when the number of sources

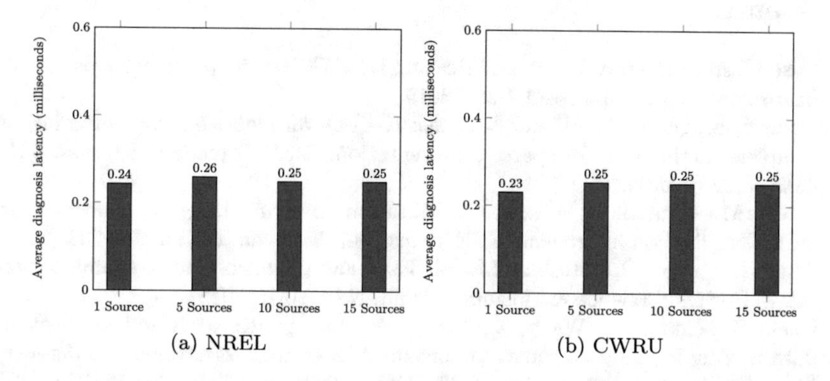

Fig. 9. Average diagnosis latency for real-time data from different number of input sources

increases. For a single source, the diagnosis latency for each signal is almost 0.24 ms, but it remains constant for α input sources, where $\alpha \in \{5, 10, 15\}$. Thus, the diagnosis latency for each data does not grow while we increase the number of input sources. As a result, the proposed system is scalable.

7 Conclusion and Future Work

In this paper, we proposed a new data-driven fault diagnosis. In particular, we proposed the model-based spectral analysis as a feature extraction method for the vibration signal. We also presented a real-time wind turbine fault diagnosis framework based on Apache Spark. The experimental results show that the extracted features help to outperform the traditional models' accuracy. The Spark-based framework can reduce the offline training time and improve the performance of the fault prediction in a real-time system. Therefore, we conclude that SAIL is scalable for real-world wind farms.

For future work, we will focus on our fault diagnosis algorithm to make it automatic and plug-&-play. Besides computational requirements in dealing with big data, another challenge emerges from a machine learning perspective. As the size of data increases, its associated variability increases. In traditional machine learning, the variability is reduced by proper feature extraction. Our intuition about the problem inspires our current feature extraction. However, big data is less intuitive. In such scenario, applying a feature learning algorithm is a possible solution. In future, we intend to apply deep neural network for feature learning in fault diagnosis of wind turbines.

Acknowledgment. This material is based upon work supported by the National Science Foundation (NSF) award number SBE-SMA-1539302, DMS-1737978. Any opinions, findings, and conclusions or recommendations expressed in this material are those of the authors and do not necessarily reflect the views of the National Science Foundation.

References

1. Case Western Reserve University Bearing Data Center. http://csegroups.case.edu/bearingdatacenter. Accessed 4 Jan 2015
2. Aharon, M., Elad, M., Bruckstein, A.: K-SVD: an algorithm for designing overcomplete dictionaries for sparse representation. IEEE Trans. Sig. Process. **54**(11), 4311–4322 (2006)
3. Amar, M., Gondal, I., Wilson, C.: Vibration spectrum imaging: a novel bearing fault classification approach. IEEE Trans. Ind. Electron. **62**(1), 494–502 (2015)
4. Chen, J., Patton, R.J.: Robust Model-Based Fault Diagnosis for Dynamic Systems, vol. 3. Springer Science & Business Media, New York (2012)
5. Chen, Z., Zhang, L., Wang, Z., Liang, W., Li, Q.: Research and application of data mining in fault diagnosis for big machines. In: International Conference on Mechatronics and Automation, 2007, ICMA 2007, pp. 3729–3734. IEEE (2007)
6. Davis, G., Mallat, S., Avellaneda, M.: Adaptive greedy approximations. Constr. Approx. **13**(1), 57–98 (1997)

7. Estabrooks, A., Jo, T., Japkowicz, N.: A multiple resampling method for learning from imbalanced data sets. Comput. Intell. **20**(1), 18–36 (2004)
8. Haddad, R.Z., Lopez, C.A., Pons-Llinares, J., Antonino-Daviu, J., Strangas, E.G.: Outer race bearing fault detection in induction machines using stator current signals. In: 2015 IEEE 13th International Conference on Industrial Informatics (INDIN), pp. 801–808. IEEE (2015)
9. Harmouche, J., Delpha, C., Diallo, D.: Improved fault diagnosis of ball bearings based on the global spectrum of vibration signals. IEEE Trans. Energy Convers. **30**(1), 376–383 (2015)
10. Hastie, T.: The Elements of Statistical Learning: Data Mining, Inference, and Prediction. Springer, New York (2009)
11. Hu, Q., He, Z., Zhang, Z., Zi, Y.: Fault diagnosis of rotating machinery based on improved wavelet package transform and SVMs ensemble. Mech. Syst. Sig. Process. **21**(2), 688–705 (2007)
12. Imani, M.B., Chandra, S., Ma, S., Khan, L., Thuraisingham, B.: Focus location extraction from political news reports with bias correction. In: 2017 IEEE International Conference on Big Data (Big Data), pp. 1956–1964. IEEE (2017)
13. Imani, M.B., Heydarzadeh, M., Khan, L., Nourani, M.: A scalable spark-based fault diagnosis platform for gearbox fault diagnosis in wind farms. In: 2017 IEEE International Conference on Information Reuse and Integration (IRI), pp. 100–107. IEEE (2017)
14. Immovilli, F., Cocconcelli, M., Bellini, A., Rubini, R.: Detection of generalized-roughness bearing fault by spectral-kurtosis energy of vibration or current signals. IEEE Trans. Ind. Electron. **56**(11), 4710–4717 (2009)
15. Kafka, A.: A high-throughput, distributed messaging system, vol. 5(1) (2014). kafka.apache.org
16. Kusiak, A., Li, W.: The prediction and diagnosis of wind turbine faults. Renew. Energy **36**(1), 16–23 (2011)
17. Kwiatkowski, D., Phillips, P.C., Schmidt, P., Shin, Y.: Testing the null hypothesis of stationarity against the alternative of a unit root: how sure are we that economic time series have a unit root? J. Econ. **54**(1–3), 159–178 (1992)
18. Lei, Y., He, Z., Zi, Y., Hu, Q.: Fault diagnosis of rotating machinery based on multiple ANFIS combination with gas. Mech. Syst. Sig. Process. **21**(5), 2280–2294 (2007)
19. Li, W., Zhang, S., Rakheja, S.: Feature denoising and nearest-farthest distance preserving projection for machine fault diagnosis. IEEE Trans. Ind. Inf. **12**(1), 393–404 (2016)
20. Mahamad, A.K., Hiyama, T.: Fault classification based artificial intelligent methods of induction motor bearing. Int. J. Innov. Comput. Inf. Control **7**(9), 5477–5494 (2011)
21. Mallat, S.: A Wavelet Tour of Signal Processing. Academic Press, Orlando (1999)
22. Meng, X., Bradley, J., Yavuz, B., Sparks, E., Venkataraman, S., Liu, D., Freeman, J., Tsai, D., Amde, M., Owen, S., et al.: MLlib: machine learning in apache spark. J. Mach. Learn. Res. **17**(34), 1–7 (2016)
23. Nelwamondo, F.V., Marwala, T., Mahola, U.: Early classifications of bearing faults using hidden markov models, gaussian mixture models, mel frequency cepstral coefficients and fractals. Int. J. Innov. Comput. Inf. Control **2**(6), 1281–1299 (2006)
24. The National Renewable Energy Laboratory (NREL): Gearbox reliability collaborative research (2009). https://www.nrel.gov/wind/grc-research.html. Accessed 1 Apr 2017

25. Pedregosa, F., Varoquaux, G., Gramfort, A., Michel, V., Thirion, B., Grisel, O., Blondel, M., Prettenhofer, P., Weiss, R., Dubourg, V., et al.: Scikit-learn: machine learning in python. J. Mach. Learn. Res. **12**, 2825–2830 (2011)
26. Pullen, A., Sawyer, S.: Global wind report. Annual market update 2014 (2014)
27. Qi, G., Tsai, W.T., Hong, Y., Wang, W., Hou, G., Zhu, Z., et al.: Fault-diagnosis for reciprocating compressors using big data. In: 2016 IEEE Second International Conference on Big Data Computing Service and Applications (BigDataService), pp. 72–81. IEEE (2016)
28. Raj, A.S., Murali, N.: Early classification of bearing faults using morphological operators and fuzzy inference. IEEE Trans. Ind. Electron. **60**(2), 567–574 (2013)
29. Riera-Guasp, M., Pineda-Sanchez, M., Perez-Cruz, J., Puche-Panadero, R., Roger-Folch, J., Antonino-Daviu, J.A.: Diagnosis of induction motor faults via gabor analysis of the current in transient regime. IEEE Trans. Instrum. Meas. **61**(6), 1583–1596 (2012)
30. Shao, Z., Wang, L., Zhang, H.: A fault line selection method for small current grounding system based on big data. In: 2016 IEEE PES Asia-Pacific Power and Energy Engineering Conference (APPEEC), pp. 2470–2474. IEEE (2016)
31. Sheng, S., Veers, P.S.: Wind turbine drivetrain condition monitoring-an overview. National Renewable Energy Laboratory (2011)
32. Shvachko, K., Kuang, H., Radia, S., Chansler, R.: The hadoop distributed file system. In: 2010 IEEE 26th Symposium on Mass Storage Systems and Technologies (MSST), pp. 1–10. IEEE (2010)
33. Solaimani, M., Gopalan, R., Khan, L., Brandt, P.T., Thuraisingham, B.: Spark-based political event coding. In: 2016 IEEE Second International Conference on Big Data Computing Service and Applications (BigDataService), pp. 14–23. IEEE (2016)
34. Apache Spark. http://spark.apache.org/
35. Stoica, P.: Spectral Analysis of Signals. Pearson/Prentice Hall, Upper Saddle River (2005)
36. Sun, W., Yang, G.A., Chen, Q., Palazoglu, A., Feng, K.: Fault diagnosis of rolling bearing based on wavelet transform and envelope spectrum correlation. J. Vib. Control **19**(6), 924–941 (2013)
37. Verma, A., Kusiak, A.: Predictive analysis of wind turbine faults: a data mining approach. In: Proceedings of the IIE Annual Conference, p. 1. Institute of Industrial and Systems Engineers (IISE) (2011)
38. Watson, S.J., Xiang, B.J., Yang, W., Tavner, P.J., Crabtree, C.J.: Condition monitoring of the power output of wind turbine generators using wavelets. IEEE Trans. Energy Convers. **25**(3), 715–721 (2010)
39. Xu, Z., Xuan, J., Shi, T., Wu, B., Hu, Y.: Application of a modified fuzzy artmap with feature-weight learning for the fault diagnosis of bearing. Expert Syst. Appl. **36**(6), 9961–9968 (2009)
40. Zappalá, D., Tavner, P.J., Crabtree, C.J., Sheng, S.: Side-band algorithm for automatic wind turbine gearbox fault detection and diagnosis. IET Renew. Power Gener. **8**(4), 380–389 (2014)
41. Zhang, Z., Verma, A., Kusiak, A.: Fault analysis and condition monitoring of the wind turbine gearbox. IEEE Trans. Energy Convers. **27**(2), 526–535 (2012)

Efficient Authentication of Approximate Record Matching for Outsourced Databases

Boxiang Dong[1,2(✉)] and Hui Wendy Wang[2]

[1] Department of Computer Science, Montclair State University, Montclair, NJ 07043, USA
dongb@montclair.edu

[2] Department of Computer Science, Stevens Institute of Technology, Hoboken, NJ 07030, USA
Hui.Wang@stevens.edu

Abstract. Cloud computing enables the outsourcing of big data analytics, where a third-party server is responsible for data management and processing. A major security concern of the outsourcing paradigm is whether the untrusted server returns correct results. In this paper, we consider *approximate record matching* in the outsourcing model. Given a target record, the service provider should return all records from the outsourced dataset that are similar to the target. Identifying approximately duplicate records in databases plays an important role in information integration and entity resolution. In this paper, we design *ALARM*, an Authentication soLution of outsourced Approximate Record Matching to verify the correctness of the result. The key idea of *ALARM* is that besides returning the similar records, the server constructs the *verification object* (*VO*) to prove their authenticity, soundness and completeness. *ALARM* consists of four authentication approaches, namely VS^2, $E\text{-}VS^2$, $G\text{-}VS^2$ and $P\text{-}VS^2$. These approaches endeavor to reduce the verification cost from different aspects. We theoretically prove the robustness and security of these approaches, and analyze the time and space complexity for each approach. We perform an extensive set of experiment on real-world datasets to demonstrate that *ALARM* can verify the record matching results with cheap cost.

Keywords: Authentication · Outsourcing
Approximate record matching · Verification object · Game theory

1 Introduction

The development of the technology to generate, collect and transmit data promotes the wide acknowledgement of the *big data era*. For example, in a single month, the US Library of Congress collects 525 terabytes of web archived data Zimmer (2015). By 2020, every human being is going to produce 1.7 megabytes

T. Bouabana-Tebibel et al. (Eds.): IEEE IRI 2017, AISC 838, pp. 119–168, 2019.
https://doi.org/10.1007/978-3-319-98056-0_6

of new data in every second Turner et al. (2014). Endowed with such abundant collection of information, big data analytics offer the promise of providing valuable insights of knowledge. Numerous data analysis techniques have been proposed, including clustering Guha and Mishra (2016); Steinbach et al. (2000), regression Draper and Smith (2014); O'Connell and Koehler (2005); Smola and Schölkopf (2004) and classification De Lathauwer et al. (2000). Among them, *approximate record matching* is the operation that searches for the records in a dataset that approximately match a given target record Hazay et al. (2007); Xiao et al. (2008). It enjoys huge importance in information integration, data cleaning and information retrieval.

Considering the massive amount of records to be matched, as well as the intrinsic complexity of similarity evaluation, it is impractical for the small- or medium-sized companies and organizations to accommodate in-house record matching support, due to the inadequacy of computational resources. A cost-effective solution to this dilemma is to outsource the record matching service to a computationally powerful service provider (server). Generally speaking, the data owner outsources a record database D to a third-party service provider (server). The server provides storage and similarity search processing as services. The search requests ask for the records in D that are similar to a given record, where the similarity is measured by a specific similarity function. Many cloud-based computational service platforms such as Amazon AWS[1] and IBM Cloud[2] are the ideal candidate host of the server.

Despite the enthralling benefits brought by outsourcing, there are several security challenges. One major concern is the *correctness* of the record matching results that are returned by the server. There exist numerous incentives for the server to cheat on the record matching result. For instance, the server may only search through a small portion of the big dataset and return incomplete result in order to save the computational efforts. It is also possible that the server intensionally manipulates the matching result to mislead the request agent due to the competitive relationship. Considering the importance of record matching result, it is imperative to check whether the service provider has performed the matching faithfully, and returned the correct results to the client.

In this paper, we only consider the record matching over categorical data, and take edit distance as the similarity/distance evaluation metric due to its popularity. Our objective is to designing a lightweight authentication framework for the client to verify the correctness of the result of outsourced record matching. We establish the concept of correctness from three perspectives: (1) *authenticity* - all the returned records are from the original dataset; (2) *soundness* - all the returned records are similar to the target record; and (3) *completeness* - all the similar records are included in the result. As the client is only equipped with limited computational resource, our main aim is to restrict the verification cost at the client side. A naive solution is to demand the client to obtain the correct result by calculating the similarity between every record in the dataset and the target record, and compare it with the result returned by the server.

[1] https://aws.amazon.com/.
[2] https://www.ibm.com/cloud/.

Obviously, the verification cost is prohibitively high, as it is equivalent to a local execution of record matching.

Existing work (e.g. Atallah et al. (2003); Dong et al. (2014); Ravikumar et al. (2004)) concentrate on a different security issue, i.e., *data privacy*, in outsourced record matching. These work protect the sensitive information by encrypting/encoding the dataset before outsourcing, and enabling the server to execute record matching on the transformed dataset. Their objective is fundamentally different from ours. Previous work on *authentication* (e.g. Li et al. (2006); Papadopoulos et al. (2011); Yang et al. (2009)) mainly focus on SQL aggregation queries and similarity search for Euclidean points. Due to the intrinsic difference in the computation task and data type, these approaches cannot be applied to authenticate record matching on categorical data. To our best knowledge, ours is the first to consider the authentication of *approximate record matching* on outsourced categorical data.

In this paper, we design *ALARM*, an Authentication soLution of outsourced Approximate Record Matching. The key idea is that besides returning the similar records, the server constructs a verification object (VO) to demonstrate the authenticity, soundness and completeness of the returned record matching results. In particular, we make the following contributions.

First, we design a lightweight authentication approach named VS^2 for approximate record matching. VS^2 is based on a new authentication tree structure called MB^{ed}-tree, which is an integration of Merkle hash tree R.C.Merkle (1980), a popularly-used authenticated data structure, and B^{ed}-tree Zhang et al. (2010), a compact index for efficient string similarity search based on edit distance. In VS^2, the VO consists of certain nodes in the MB^{ed}-tree. The indexing functionality of the MB^{ed}-tree facilitates the efficient VO verification procedure at the client side.

Second, we design an optimized approach named E-VS^2. E-VS^2 applies a similarity-preserving embedding function to map records to the Euclidean space in the way that similar records are mapped to close Euclidean points. The VO is constructed from both the MB^{ed}-tree and the embedded Euclidean space. Compared with VS^2, E-VS^2 saves the verification cost by replacing a large amount of expensive string edit distance calculation with a small number of cheap Euclidean distance computation.

Third, we propose the G-VS^2 approach to further reduce the verification cost at the client side. G-VS^2 generates the *gram counting vector* (GC-vector) of every record, where the distance calculation over GC-vectors is much more efficient than string edit distance. For those records whose similarity/dissimilarity to the target can be derived from the GC-vectors, G-VS^2 includes their GC-vectors in the VO. Compared with VS^2 and E-VS^2, G-VS^2 saves the verification cost at the client side.

Fourth, by involving the server and the client into a strategic game Morris (2012), we design another approach named P-VS^2. We carefully design the strategic game in the way that by only requesting the client to calculate the edit distance for a small fraction of records, the server always returns the correct result in order to gain the most payoff.

Fifth, we formally prove that any integrity violation of record matching results can be caught by our four authentication approaches. We also prove these approaches are secure under the standard cryptographic assumption of collision resistant hash functions and intractability of integer factorization problem. Moreover, we analyze the time and space complexity of each of the four approaches.

Last but not least, we conduct an extensive set of experiments on real-world datasets. The empirical results demonstrate the efficiency of our authentication approaches. Compared with the baseline, VS^2 and $G\text{-}VS^2$ can reduce the verification cost at the client side by 53% and 96% respectively. $P\text{-}VS^2$ can verify the record matching result over 1 million records with only 216 edit distance calculations.

The rest of the paper is organized as follows. Section 2 discusses the preliminaries. Section 3 defines the problem. Section 4 presents our four authentications approaches in detail. Section 5 analyzes the time and space complexity of these approaches. The experiment results are shown in Sect. 6. Sections 7 discusses the related work. Section 8 concludes the paper.

2 Preliminaries

2.1 Record Similarity Measurement

Record matching is a fundamental problem in many research areas, e.g., information integration, database joins, and more. In general, the evaluation that whether two records match is mainly based on the string similarity of the attribute values. There are a number of string similarity functions, e.g., Hamming distance, n-grams, and edit distance (see Koudas et al. (2006) for a good tutorial). In this paper, we mainly consider *edit distance*, one of the most popular string similarity measurements that have been used in a wide spectrum of applications. Informally, the edit distance of two strings s_1 and s_2, denoted as $DST(s_1, s_2)$, measures the minimum number of insertion, deletion and substitution operations to transform s_1 to s_2. We say two strings s_1 and s_2 are *similar*, denoted as $s_1 \approx s_2$, if $DST(s_1, s_2) \leq \theta$, where θ is a user-specified similarity threshold. Otherwise, we say s_1 and s_2 are *dissimilar* (denoted as $s_1 \not\approx s_2$). Without loss of generality, we assume that the dataset D only contains a single attribute. Thus, in the following sections, we use record and string interchangeably. Our methods can be easily adapted to the datasets that have multiple attributes. Given a dataset D, a target string s_q, and a similarity threshold θ, the *record matching* problem is to find all strings in D that are similar to s_q, i.e., $M = \{s \in D | DST(s_q, s) \leq \theta\}$.

2.2 Mapping Strings to Euclidean Space

Given two strings s_1 and s_2, normally the complexity of computing edit distance is $O(|s_1||s_2|)$, where $|s_1|$ and $|s_2|$ are the lengths of s_1 and s_2. One way to reduce the complexity of similarity measurement is to map the records into a multi-dimensional Euclidean space, such that the similar strings are mapped to close Euclidean points. The main reason of the embedding is that the computation of Euclidean distance is much cheaper than string edit distance. A few string embedding techniques (e.g., Faloutsos and Lin (1995); Hjaltason and Samet (2003); Jin et al. (2003)) exist in the literature. These algorithms have different properties in terms of their efficiency and distortion rate (See Hjaltason and Samet (2000) for a good survey). In this paper, we consider an important property named *contractiveness* property of the embedding methods, which requires that for any pair of strings (s_i, s_j) and their embedded Euclidean points (p_i, p_j), $dst^E(p_i, p_j) \leq DST(s_i, s_j)$, where $dst^E()$ and $DST()$ are the distance function in the Euclidean space and string space respectively. In this paper, we use $dst^E()$ and $DST()$ to denote the Euclidean distance and edit distance. In this paper, we use the *SparseMap* method Hjaltason and Samet (2003) for the string embedding. SparseMap preserves the contractiveness property. We will show how to leverage the contractiveness property to improve the verification performance in Sect. 4.3. Note that the embedding methods may introduce false positives, i.e. the embedding points of dissimilar strings may become close in the Euclidean space.

2.3 Approximate String Matching Based on Q-Grams

Given a constant q, a *q-gram* of a string s is a consecutive sequence of q characters in s. For any string s, we can construct its q-gram set $Q(s)$ by sliding a window of size q through s. To embody q-grams from the beginning and end of s, we extend s by prefixing and suffixing it with $q-1$ special characters #, where # is a character outside the alphabet Σ of the strings. Example 1 displays an example of q-gram construction.

Example 1. Consider the string $s = JOHN$, and $q = 2$. To construct the set of 2-grams of s, we first insert # to the beginning and end of s, and obtain #JOHN#. Then we extract the set of 2-grams as $Q(s) = \{\#J, JO, OH, HN, J\#\}$.

An interesting property is that similar strings with small edit distance also have a large overlap in their q-grams. Previous work Xiao et al. (2008) prove that for any pair of strings s_1 and s_2 such that $DST(s_1, s_2) = 1$, $Q(s_1)$ and $Q(s_2)$ must differ with at most q elements. Informally, this is because the alternation of a single character can leads to the distortion of at most q grams. Xiao et al. (2008) presents the following lemma between the intersection of q-grams and edit distance between strings.

Lemma 1. *Given any pair of strings s_1 and s_2, let $|s_1|$ and $|s_2|$ be their length. For any constant q, let $|Q(s_1) \cap Q(s_2)|$ be the number of common q-grams of s_1 and s_2. It must be true that*

$$|Q(s_1) \cap Q(s_2)| \geq max(|s_1, |s_2||) - 1 - (DST(s_1, s_2) - 1) \times q.$$

Compared to string edit distance computation, it is much for efficient to calculate the cardinality of the intersection of q-grams, i.e. $|Q(s_1) \cap Q(s_2)|$. Following Lemma 1, quite a few papers Sutinen and Tarhio (1995); Ukkonen (1992); Xiao et al. (2008) optimizes the efficiency of approximate string matching based on q-grams.

2.4 Authenticated Data Structure

To enable the client to authenticate the correctness of mining results, the server returns the results along with some supplementary information that permits result verification. Normally the supplementary information takes the format of *verification object* (VO). In the literature, VO generation is usually performed by an authenticated data structure (e.g., Li et al. (2006); Papadopoulos et al. (2011); Yang et al. (2009)). One of the popular authenticated data structures is *Merkle tree* R.C.Merkle (1980). In particular, a Merkle tree is a tree T in which each leaf node N stores the digest of a record r: $h_N = h(r)$, where $h()$ is a one-way, collision-resistant hash function (e.g. SHA-1). For each non-leaf node N of T, it is assigned the value $h_N = h(h_{C_1}||\dots||h_{C_k})$, where $C_1, \dots, C_k$ are the children of N. The root signature sig is generated by signing the digest h_{root} of the root node using the private key of a trusted party (e.g., the data owner). The VO enables the client to re-construct the root hash value.

In the classic security definition Papamanthou and Tamassia (2007), an authentication approach is secure if no probabilistic polynomial-time adversary $\mathcal{A}$ with oracle access to the VO generation routine, can produce an incorrect result M' and VO' that pass the verification process with non-negligible probability to the security parameter λ, i.e.,

$$Pr[((M', VO') \leftarrow \mathcal{A}(F)) \wedge (1 \leftarrow Verify(M', VO'))] = \nu(\lambda),$$

where F is the VO construction function, and $\nu(\lambda)$ is a negligible function to λ. Informally, this definition limits the likelihood that an adversary can forge a counterfeit VO of an incorrect result and pass the verification.

2.5 Condensed-RSA

Although effective, traditional *RSA* incurs significant communication cost to authenticate a sequence of messages. Specifically, to demonstrate the authenticity of n messages $s_1, \dots, s_n$, n signatures $\sigma_1, \dots, \sigma_n$ needs to be transmitted from the prover to the verifier. To overcome the drawback, Mykletun et al. (2006) proposed a signature aggregation scheme named *Condensed-RSA*, which is to compress a set of RSA signatures Rivest et al. (1978) into a single signature. In specific, Condensed-RSA consists of three probabilistic polynomial-time algorithms $(Gen, Sign, Vrfy)$ that are defined as follows.

- $(pk, sk) \leftarrow Gen(1^\lambda)$: on input 1^λ, generate two λ-bits random prime numbers p and q, and obtain $N = pq$. Find a pair of integers e and d that satisfy $ed = 1 \ mod \ \phi(N)$. Output the public key $pk =< N, e >$ and the private key $sk =< N, d >$.

- $\sigma_{1,n} \leftarrow Sign(sk, \{m_1, \ldots, m_n\})$: given a set of messages $\{m_1, \ldots, m_n\}$, compute the individual signatures $\sigma_i = H(m_i)^d \ mod \ \phi(N)$, where H is a full-domain cryptographic hash function that converts a message into a value in Z_N*. Then compute the aggregate signature

$$\sigma_{1,n} = \Pi_{i=1}^{n} \sigma_i \ (mod \ N). \tag{1}$$

- $0/1 \leftarrow Vrfy(pk, \{m_1, \ldots, m_n\}, \sigma_1^n)$: on input a sequence of messages $\{m_1, \ldots, m_n\}$ and an aggregate signature σ_1^n), output 1 if and only if

$$(\sigma_{1,n})^e \stackrel{?}{\equiv} \Pi_{i=1}^{n} H(s_i) \ (mod \ N). \tag{2}$$

It has been proved that the Condensed-RSA signature scheme is existentially unforgeable under an adaptive chosen-message attack for all probabilistic polynomial-time adversaries. Compared with the verification approach based on traditional RSA Rivest et al. (1978), the Condensed-RSA approach reduces the number of modular exponentiations at the verifier side by an order of magnitude, and saves the communication overhead between the prover and the verifier from n to 1 RSA signatures.

2.6 B^{ed}-Tree for String Similarity Search

A number of compact data structures (e.g., Arasu et al. (2006); Chaudhuri et al. (2003); Li et al. (2008)) are designed to handle edit distance based similarity measurement. In this paper, we consider B^{ed}-*tree* Zhang et al. (2010) due to its support for external memory and dynamic data updates. B^{ed}-tree is a B^+-tree based index structure that can handle arbitrary edit distance thresholds. The tree is built upon a *string ordering* scheme which is a mapping function φ to map each string to an integer value. To simplify notation, we say that $s \in [s_i, s_j]$ if $\varphi(s_i) \leq \varphi(s) \leq \varphi(s_j)$. Based on the string ordering, each B^{ed}-tree node N is associated with a string range $[N_b, N_e]$. Each leaf node contains $f \geq 1$ strings $\{s_1, \ldots, s_f\}$, where $s_i \in [N_b, N_e]$, for each $i \in [1, f]$. Each intermediate node N (with range $[N_b, N_e]$) contains multiple children nodes, where for each child, its range $[N'_b, N'_e] \subseteq [N_b, N_e]$. We say these strings that stored in the sub-tree rooted at N as the strings that are *covered* by N.

We use $DST_{min}(s_q, N)$ to denote the *minimal* edit distance between a string s_q and any string s that is covered by a B^{ed}-tree node N. A nice property of the B^{ed}-tree is that, for any string s_q and node N, the string ordering φ enables to compute $DST_{min}(s_q, N)$ efficiently by computing $DST(s_q, N_b)$ and $DST(s_q, N_e)$ only, where N_b, N_e refer to the string range values of N. Based on this, we define the B^{ed}-*tree candidate node*.

Definition 1. *Given a string s_q and a similarity threshold θ, a B^{ed}-tree node N is a* candidate *if $DST_{min}(s_q, N) \leq \theta$. Otherwise, N is a* non-candidate. ∎

B^{ed}-tree has an important monotone property.

Property 21. (Monotone Property of B^{ed}-tree: Given a node N_i in the B^{ed}-tree and any child node N_j of N_i, for any string s_q, it must be true $DST_{min}(s_q, N_i) \leq DST_{min}(s_q, N_j)$.

Therefore, for any non-candidate node, all of its children must be non-candidates. For any given string s_q, the string similarity search algorithm starts from the root of the B^{ed}-tree, and iteratively visits the candidate nodes, until all candidate nodes are visited. The algorithm does not visit non-candidate nodes as well as their descendants. This monotone property enables early termination of search on the branches that contain non-candidate nodes.

An important note is that for each candidate node, some of its covered strings may still be dissimilar to s_q. Therefore, given a string s_q and a candidate B^{ed}-tree node N, it is necessary to compute $DST(s_q, s)$, for each s that is covered by N.

2.7 Game Theory

A strategic game involves a set of players, where each player can have multiple actions to take. For each player, any action is associated with a payoff Morris (2012). It is worth noting that the payoff for a player may be affected by the decisions of the other players. All the players in the game are *rational choice-makers*. To be precise, every player picks the action that maximizes his/her payoff. A *Nash equilibrium* in a game is a profile of actions for all players such that no single player can gain more benefits by switching to another action. In other words, a Nash equilibrium encompasses a *stable* state of the players in a game, i.e., once reached, all players would adhere to their decisions eternally.

3 Problem Formulation

3.1 System Model

We consider the outsourcing model that involves three parties - a data owner who possesses a dataset D that contains n records, the user (client) who requests for approximate record matching services on D, and a third-party service provider (server) that executes the record matching services on D. The client can be the data owner or an authorized party. Initially, the data owner outsources D to the server. The server provides storage and record matching as services. Upon receiving a record matching request (s_q, θ) from any client, the server searches for the matching result M^S. Additionally, to facilitate result authentication, the server constructs a *verification object* VO to demonstrate the correctness of M^S. The server transmits both M^S and VO to the client. The client checks the correctness of M^S via VO. Given the fact that the client may not possess D, we require that the availability of D is not necessary for the authentication procedure.

3.2 Threat Model

We assume that the third-party server is not fully trusted as it could be compromised by the attacker (either inside or outside). The server may alter the received dataset D or return any matching result that does not exist in D. It may also tamper with the matching results. For instance, the server may return *incomplete* results that omit some legitimate similar records in the matching results Pang and Mouratidis (2008). Given the importance of the similarity search results, it is vital to provide efficient authentication mechanisms that enable the client to verify the correctness of the search results. Note that we do not consider privacy protection for the outsourced data. This issue can be addressed by privacy-preserving record linkage Dong et al. (2014) and is beyond the scope of this paper.

3.3 Verification Goal of Result Integrity

Given a dataset D, a target string s_q, and the similarity threshold θ, let M^S be the record matching result returned by the server. To catch the cheating on the matching results, we formally define the integrity of the record matching result from three perspectives:

- **Result authenticity.** All the strings in M^S must be present in the original dataset D, and be intact. In other words, there exists no string s such that $s \in M^S$, but $s \notin D$.
- **Result soundness.** All strings in M^S must be similar to the target s_q. In other words, for every string $s \in M^S$, it must be true that $DST(s, s_q) \leq \theta$.
- **Result completeness.** All strings that are similar to s_q must be included in M^S, i.e., for every string $s \in D$ such that $DST(s, s_q) \leq \theta$, it must be true that $s \in M^S$.

In this paper, we design an efficient and secure verification system named *ALARM* for the client to verify the authenticity, soundness and completeness of matching results M^S.

4 *ALARM* Authentication System

In this section, we first introduce the verification framework of *ALARM* in Sect. 4.1. Then we discuss our four authentication approaches named VS^2, *E-*VS^2, *G-*VS^2 and *P-*VS^2 in Sects. 4.2 to 4.5 respectively.

4.1 System Overview

We design an efficient authentication system named *ALARM* to check the correctness of the outsourced record matching computations. *ALARM* consists of three phases, which are displayed in Fig. 1.

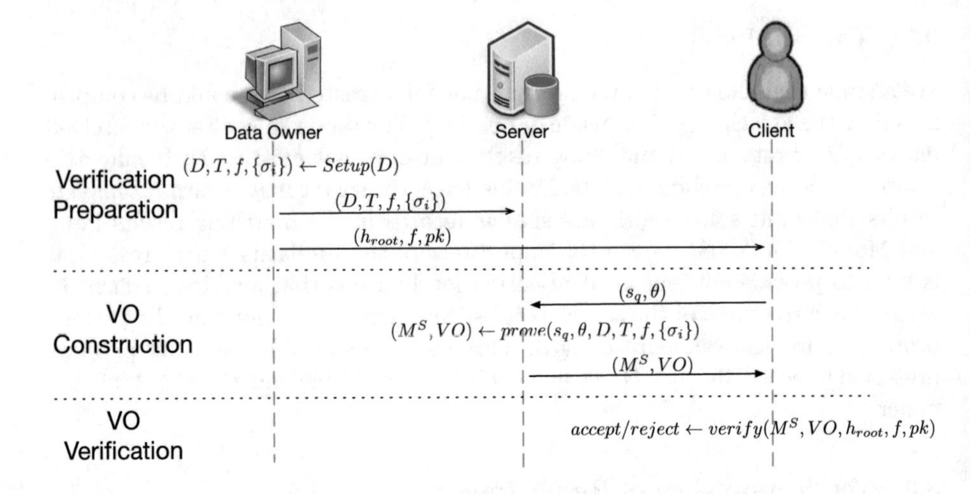

Fig. 1. An overview of the three phases of *ALARM*

- **Verification preparation phase.** The data owner constructs an authenticated tree structure T and RSA signatures $\{\sigma_i\}$ from the dataset D, and takes the root hash value h_{root} of the authenticated data structure as the digest. Moreover, the data owner generates the embedding function f that transforms the strings to Euclidean points. The data owner sends the dataset D, the ADS T, the signatures $\{\sigma_i\}$ and f to the server. To reduce the network bandwidth consumption and the store overhead at the client side, the data owner only needs to publish the root hash value h_{root}, the embedding function f, and the RSA public key $pk = (N, e)$ to the legitimate clients.
- **VO construction phase.** Upon receiving a record matching request (s_q, θ) from the client, the server executes the computation to obtain the result M^S. In order to demonstrate the correctness of M^S, the server also constructs the verification object VO of M^S from the ADS T, the embedded Euclidean points of D, and the RSA signatures. After that, the server sends both M^S and VO to the client to prove the correctness of the result.
- **VO verification phase.** The client inspects the correctness of M^S by checking the VO against the auxiliary information (i.e., h_{root}, f, and $pk = (N, e)$) obtained from the data owner.

It is worth noting that even though we only consider string similarity matching in this paper, the authentication approaches that we propose can be easily adapted to support numerical values by switching the underlying indexing structure to R-tree Beckmann et al. (1990). Besides, the preparation phase is only a one-time setup at the data owner side. Once the ADS, the embedding function f and the RSA signatures are produced, they can be readily utilized for all the matching requests. Therefore, the one-time verification preparation cost at the data owner side can be amortized over the result verification of all future record matching requests.

4.2 Basic Approach: VS^2

We design a verification method of similarity search (VS^2) approach. The key component of VS^2 is the authenticated data structure (ADS), which is the authenticated version of MB^{ed}-tree. In the *verification preparation* phase, the data owner constructs the authenticated data structure T of the dataset D. In the *VO construction* phase, the server executes the approximate matching on D to find similar strings for s_q, and constructs the *verification object* (VO) of the search results M^S by traversing T. The server returns both M^S and VO to the client. In the *VO verification* phase, the client verifies the integrity of M^S by leveraging VO and h_{root}. Next we explain the details of these three phases.

4.2.1 Verification Preparation

We design a new authenticated data structure named the *Merkle B^{ed}-tree* (MB^{ed}-tree). The MB^{ed}-tree is constructed on top of the B^{ed}-tree by assigning a digest to each B^{ed}-tree node. Every MB^{ed}-node include a certain number of entries and a triplet (N_b, N_e, h_N). Let f be the fanout, i.e., the number of entries in a node. Then N_b and N_e correspond to the string range values associated with N, and h_N is the digest value computed as $h_N = H(H(N_b)||H(N_e)||h^{1\to f})$, where $h^{1\to f} = H(h_{C_1}||\dots||h_{C_f})$, with $C_1, \dots, C_f$ being the children of N, and H denoting the collision resistant hash function. For any leaf node N, every entry of N stores a pair (s, p), where s is a string covered by N, and p is the pointer to the disk block that stores s. If N is an internal node, each entry contains a pointer to one of its children nodes.

The digests of the MB^{ed}-tree T can be constructed in the bottom-up fashion, starting from the leaf nodes. After all nodes of T are associated with the digest values, the data owner keeps the root hash value h_{root} locally. Note that it is possible that the data owner signs the root with her private key by using a public-key cryptosystem (e.g., RSA) to defend against the *man-in-the-middle attack* Merkle (1978). However, it is beyond the scope of this paper. An example of the MB^{ed}-tree structure is presented in Fig. 2. The data owner sends both D and T to the server. The data owner only sends h_{root} to any client who requests it for authentication purpose.

Following Comer (1979); Li et al. (2006), we assume that each node of the MB^{ed}-tree occupies a disk page. For the MB^{ed}-tree T, each entry in the leaf node occupies $|s| + |p|$ space, where $|p|$ is the size of a pointer, and $|s|$ is the maximum length of a string value. The triple (N_b, N_e, h_N) takes the space of $2|s| + |h|$, where $|h|$ is the size of a hash value. Therefore, a leaf node can have $f_1 = [\frac{P-2|s|-|h|}{|p|+|s|}]$ entries at most, where P is the page size. Given n unique strings in the dataset, there are $[\frac{n}{f_1}]$ leaf nodes in T. Similarly, for the internal nodes, each entry takes the space of $|p|$. Thus each internal node can have at most $f_2 = [\frac{P-2|s|-|h|}{|p|}]$ entries (i.e., $[\frac{P-2|s|-|h|}{|p|}]$ children nodes). Therefore, the height ht of T $ht \geq log_{f_2}[\frac{n}{f_1}]$. To support efficient incremental data updates, we can simply set smaller f_1 and f_2 values to accommodate space for new strings to be inserted. The insertion of a string only triggers the update of digests of the tree nodes in the path from the specific leaf node to the root, which incurs $O(ht)$ complexity.

4.2.2 VO Construction

For each target string s_q, the server constructs a VO from the MB^{ed}-tree T to show that the matching result M^S is both sound and complete.

First, we define M-strings and *false hits*. Given a target string s_q and a similarity threshold θ, any string $s \in D$ can be classified as one of the following two types:

- M-**string** if $s \approx s_q$, i.e., $DST(s, s_q) \leq \theta$; or
- **False hit** if $s \not\approx s_q$, i.e., $DST(s, s_q) > \theta$.

We denote the set of *false hits* with F. In other words, $F = \{s | s \in D, s_q \not\approx s\}$. Intuitively, to verify that M^S is sound and complete, the VO includes both similar strings M^S and false hits F. Apparently including all false hits may lead to a large VO, and thus high network communication cost and the verification cost at the client side. Therefore, we aim to reduce the verification cost incurred by the large number of strings in F.

Before we explain how to reduce VO size, we first define C-strings and NC-strings. Apparently, each false hit string is covered by a leaf node of the MB^{ed}-tree T. Based on whether a leaf node in MB^{ed}-tree is a candidate (Definition 1 in Sect. 2.6), we can classify any false hit s into one of the two types:

- C-**string** (in short for *candidate string*): if s is covered by *candidate* leaf nodes; and
- NC-**string** (in short for *non-candidate string*): if s is covered by *non-candidate* leaf nodes.

Our key idea to reduce verification cost is to use *representatives* of NC-strings in VO instead of individual NC-strings. The representatives of NC-strings take the format of *maximal false hit*-nodes (MF-nodes).

Definition 2. *Given a target string s_q and a threshold θ, a MB^{ed}-tree node N is a MF-node if N is a* non-candidate *node, and N's parent is a* candidate *node.*

Note that the MF-nodes can be leaf nodes. Apparently, all strings covered by the MF-nodes must be NC-strings. Conversely, each NC-string must be covered by a MF-node. Furthermore, MF-nodes are disjoint (i.e., no two MF-nodes cover the same string). Therefore, instead of including individual NC-strings into the VO, the MF-nodes are included. Consequently, instead of calculating the edit distance for individual NC-strings, the client only needs to calculate $DST_{min}(s_q, N)$ for the MF-nodes in VO verification phase. The advantage is that the number of MF-nodes is much smaller than the number of NC-strings. Our empirical study shows that a single MF-node can cover 495 NC-strings. Therefore, we can effectively reduce the VO verification cost. Now we are ready to define the VO.

Definition 3. *Given a dataset D and a target string s_q, let M^S be the returned similar strings of s_q. Let T be the MB^{ed}-tree of D. Then the VO of s_q consists of:*

(i) string s, for each M-string or C-string; and
(ii) a pair $([N_b, N_e], h^{1\to f})$ for each MF-node N, where $[N_b, N_e]$ denote the string range associated with N, and $h^{1\to f} = h(h_{C_1}||\cdots||h_{C_f})$, with $C_1, \ldots, C_f$ being the children of N.

Furthermore, in VO, a pair of square bracket is added around the strings that are located in the same node of T.

Intuitively, in VO, the M-strings and C-strings are present in the original string format, while NC-strings are represented by the MF-nodes (i.e., in the format of $([N_b, N_e], h^{1\to f})$).

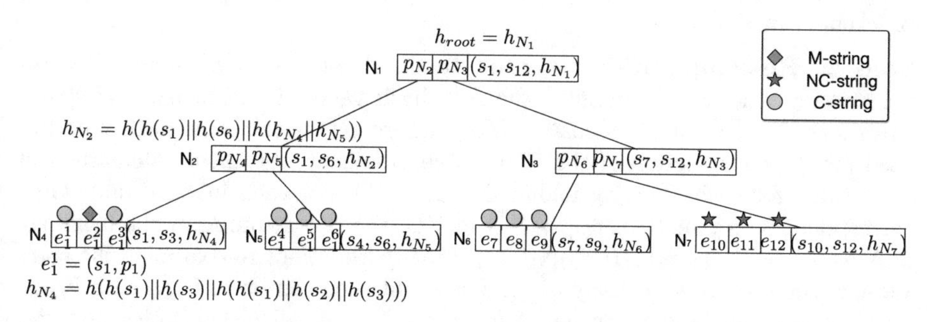

Fig. 2. An example of Merkle B^{ed} tree

Example 2. Consider the MB^{ed}-tree T in Fig. 2, and the string s_q to construct VO. Assume the only M-string is s_2. Also assume that node N_7 of T is the only MF-node. Then the NC-strings are s_{10}, s_{11}, s_{12}, and C-strings are $s_1, s_3, s_4, \ldots s_9$. Therefore, the VO is

$$VO = \{(((s_1, s_2, s_3), (s_4, s_5, s_6)), ((s_7, s_8, s_9), ([s_{10}, s_{12}], h^{10\to 12})))\}, where\ h^{10\to 12} = H(H(s_{10})||H(s_{11})||H(s_{12})).$$

For each MF-node, we do not require that $H(N_b)$ and $H(N_e)$ appear in VO. Instead, the server must include both N_b and N_e in VO. This is to prevent the server from cheating on the MF-nodes by including incorrect $[N_b, N_e]$ in VO. More details of the robustness analysis of our authentication procedure can be found in Sect. 4.2.4.

4.2.3 VO Verification

For a given target string s_q, the server returns VO together with the matching set M^S to the client. The verification procedure consists of three steps. In Step

1, the client re-constructs the MB^{ed}-tree from VO. In Step 2, the client re-computes the root hash value h'_{root}, and compares h'_{root} with the hash value h_{root} received from the data owner. In Step 3, the client re-computes the edit distance between s_q and a *subset* of strings in VO. Next, we explain the details.

Step 1: Re-construction of MB^{ed}-tree. First, the client sorts the strings and string ranges (in the format of $[N_b, N_e]$) in VO by their mapping values according to the string ordering scheme. String s is put ahead of the range $[N_b, N_e]$ if $s < N_b$. It returns a total order of strings and string ranges. If there exists any two ranges $[N_b, N_e]$ and $[N'_b, N'_e]$ that overlap, the client concludes that the VO is not correct. If there exists a string $s \in M^S$ and a range $[N_b, N_e] \in VO$ such that $s \in [N_b, N_e]$, the client concludes that M^S is not sound, as s indeed is a dissimilar string (i.e., it is included in a non-candidate node). Second, the client maps each string $s \in M^S$ to an entry in a leaf node in T, and each pair $([N_b, N_e], h_N) \in VO$ to an internal node in T. The client re-constructs the parent-children relationships between these nodes by following the matching brackets () in VO.

Step 2: Re-computation of root hash. After the MB^{ed}-tree T is re-constructed, the client computes the root hash value of T. For each M-string value $s \in M^S$, the client calculates $H(s)$, where $H()$ is the same hash function used for the construction of the MB^{ed}-tree. Similarly, the client calculates the hash value for each C-string included in the VO. For each internal node that corresponds to a pair $([N_b, N_e], h^{1 \to f})$ in VO, the client computes the hash h_N of N as $h_N = H(H(N_b)||H(N_e)||h^{1 \to f})$. Finally, the client re-computes the hash value of the root node, namely h'_{root}. The client then compares h'_{root} with h_{root}, which is sent by the data owner. If $h'_{root} \neq h_{root}$, the client concludes that the server's results are not authentic.

Step 3: Re-computation of necessary edit distance. First, for each string $s \in M^S$, the client re-computes the edit distance $DST(s_q, s)$, and verifies whether $DST(s_q, s) \leq \theta$. If all strings $s \in M^S$ pass the verification, then the client concludes that M^S is *sound*. Second, for each C-string $s \in VO$ (i.e., those strings appear in VO but not M^S), the client verifies whether $DST(s_q, s) > \theta$. If it is not (i.e., s is a similar string indeed), the client concludes that the server fails the completeness verification. Third, for each range $[N_b, N_e] \in VO$, the client verifies whether $DST_{min}(s_q, N) > \theta$, where N is the corresponding MB^{ed}-tree node associated with the range $[N_b, N_e]$. If it is not (i.e., node N is indeed a candidate node), the client concludes that the server fails the completeness verification.

Example 3. Consider the MB^{ed}-tree in Fig. 2 as an example, and the target string s_q. Assume the similar strings $M^S = \{s_2\}$. Consider the VO shown in Example 2. The C-strings are $s_1, s_3, s_4, \ldots s_9$. After the client re-constructs the MB^{ed}-tree, It computes the root hash value h'_{root} from VO by executing the following procedures:

(1) recover N_4 $([s_1, s_3], h_{N_4})$, where $h_{N_4} = H(H(s_1)||H(s_3)||h^{1 \to 3})$, $h^{1 \to 3} = H(H(s_1)||H(s_2)||H(s_3))$, and s_1, s_2 and s_3 are from the VO;

(2) recover N_5 $([s_4, s_6], h_{N_5})$ *and* N_6 $([s_7, s_9], h_{N_6})$ *in the same manner;*
(3) recover N_2 $([s_1, s_6], h_{N_2})$ *from* N_4 *and* N_5*;*
(4) recover N_3 $([s_7, s_{12}, h_{N_3}])$ *from* N_6 *and* N_7*, where* N_7 *is the MF-node included in* VO*; and*
(5) re-compute $h'_{root} = H(H(s_1)||H(s_2)||h^{1\to 12})$ *from* N_2 *and* N_3*.*

After that, the client compares it against h_{root}*. It also performs the following distance computations:*

(1) for $M^S = \{s_2\}$*, compute* $DST(s_q, s_2)$*;*
(2) for the C*-strings* $s_1, s_3, s_4, \ldots, s_9$*, compute the edit distance between them and* s_q*; and*
(3) for the pair $([s_{10}, s_{12}], h^{10\to 12}) \in VO$*, compute* $DST_{min}(s_q, N_7)$*.*

Compared with the record matching locally which requires 12 edit distance calculations, VS^2 *only computes 10 edit distances.*

4.2.4 Robustness Analysis

Given a target string s_q and a similarity threshold θ, let M be the similar strings of s_q. Let M^S be the similar strings of s_q returned by the server. An untrusted server may perform the following cheating behaviors to generate M^S: (1) *authenticity violation*: include at least one string s' that does no exist in the original dataset D in M^S ; (2) *soundness violation*: the server returns $M^S = M \cup \{s'\}$, where $s' \in D$, but $s' \not\approx s_q$; and (3) *completeness violation*: the server returns $M^S = M - \{s'\}$ for some string $s' \in D$ and $s' \approx s_q$.

Authenticity. The authenticity violation can be easily caught by the authentication procedure, as the hash values of the tampered strings are not the same as the original strings. This leads to that the root hash value of MB^{ed}-tree reconstructed by the client different from the root value of original dataset. The client can catch the tampered values by Step 2 of the authentication procedure.

Soundness. The server may violate the soundness requirement by including a string $s' \in D$, $s' \not\approx s_q$ into M^S. The soundness violation can be easily detected in Step 3 of verification. This is because by calculating the edit distance for each string in M^S, the client must be able to find that $DST(s', s_q) > \theta$.

Completeness. The server may return incomplete result by excluding a string $s' \in D$, $s' \approx s_q$ from the result. In other words, $M^S = M - \{s'\}$. In the VO, it is possible that the server excludes s', or claims s' as a C-string or NC-string.

Case 1: s' is excluded from VO. The client detects it in Step 2 This is because with a missing string in VO, it must be true $h'_{root} \neq h_{root}$.

Case 2: s' is treated as a C-string. The client catches the incomplete result in Step 3. By calculating the edit distance for every C-string in VO, the client finds that $DST(s', s_q) \leq \theta$.

Case 3: The server claims s' as an NC-string by including s' into a MF-node N. Then we have the following theorem for the robustness analysis.

Theorem 1. Given a target string s_q and a MF-node N of range $[N_b, N_e]$, including any string s' into N such that $s' \approx s_q$ will change N to be a candidate node.

Proof. The proof is straightforward. According to the definition of the function $DST_{min}(s_q, N)$ in Sect. 2.6, we have $DST_{min}(s_q, N) \leq DST(s_q, s)$ for any string s that is covered by N. As $DST(s_q, s') \leq \theta$, it must be true that $DST_{min}(s_q, N) \leq \theta$. Therefore, N becomes a candidate node after inserting s' into N. ∎

Theorem 1 guarantees that the client can catch the incomplete result in Step 3, as in the process of calculating the minimum distance for every MF-node in VO, the client finds at least one candidate node.

4.2.5 Security Analysis

In this section, we demonstrate the security of VS^2, i.e., no probabilistic polynomial-time adversary with oracle access to the VO generation routine is able to produce a counterfeit VO that enables an incorrect record matching result M^{S}' to pass the verification process. We defend this claim based on the standard cryptographic assumption of collision resistant hash functions.

Outline of Security. Assume that given a sequence of t record matching requests $\{(s_1, \theta_1), \ldots, (s_t, \theta_t)\}$ and the proof $\{\pi_1, \ldots, \pi_t\}$ generated by the VO construction routine, for a new request (s_q, θ), the adversary $\mathcal{A}$ produces incorrect M' and π' that successfully escapes the 3-step verification process (Sect. 4.2.3). This implies that (1) for any $s \in M$', $DST(s_q, s) \leq \theta$; (2) for any C-string $s \in \pi$', $DST(s_q, s) > \theta$; (3) for any MF-node $N \in \pi$', $DST_{min}(s_q, N) > \theta$; and (4) the root hash value re-calculated from π' matches the original one, i.e., $h'_{root} = h_{root}$. Let π be the proof constructed by following the VO generation routine correctly. According to Miller et al. (2014), there must exist at least one collision between π and π'. If the first collision occurs at a leaf entry, i.e., the adversary $\mathcal{A}$ finds a pair of strings $s_1 \neq s_2$ s.t. $H(s_1) = H(s_2)$. Otherwise, the first collision takes place at an internal node, i.e., the adversary $\mathcal{A}$ comes up with a pair of internal nodes N_1 and N_2 s.t. $h_1^{1\to f} \neq h_2^{1\to f}$ and $h_{N_1} = h_{N_2}$.

Following Mykletun et al. (2006), we can create an adversary $\mathcal{B}$ that breaks the collision resistant hash function $H(\cdot)$. $\mathcal{B}$ proceeds in the following manner:

(1) Generates t record matching requests $\{(s_1, \theta_1), \ldots, (s_t, \theta_t)\}$;
(2) Sends these requests to $\mathcal{A}$;
(3) Let the forged result and VO produced by $\mathcal{A}$ be M' and π';
(4) Produces the correct result and VO, i.e., M and π;
(5) Outputs (s_1, s_2) (if the first collision between π and π' occurs at a leaf entry) or $(H(N_{b1})||H(N_{e1})||h_1^{1\to f}, H(N_{b2})||H(N_{e2})||h_2^{1\to f})$ (if the first collision occurs at a MB^{ed}-node).

Obviously, the pair of messages output by $\mathcal{B}$ result in a collision to the hash function $H(\cdot)$. If the probability that $\mathcal{A}$ can successfully escapes the verification process is non-negligible to λ, so is the probability that $\mathcal{B}$ breaks the hash function $H(\cdot)$. This contradicts the standard cryptographic assumption of collision resistant hash functions. Therefore, we successfully prove the security of the VS^2 authentication approach.

4.3 String Embedding Authentication: $E\text{-}VS^2$

One weakness of VS^2 is that if there exists a significant number of C-strings, the verification cost might be expensive, as the client has to compute the string edit distance between the target record and the C-strings. Our goal is to further reduce the VO verification cost at the client side by decreasing the amount of edit distance computation. We observe that although C-strings are not similar to the target record, they may be similar to each other. Therefore, we design $E\text{-}VS^2$, a computation-efficient method on top of VS^2. The key idea of $E\text{-}VS^2$ is to construct a set of *representatives* of C-strings based on their similarity, and only include the representatives of C-strings in VO. To construct the representatives of C-strings, we first apply a similarity-preserving embedding function on C-strings, and transform them into the Euclidean space, so that the similar records are mapped to the close points in the Euclidean space. Then C-strings are organized into a small number of groups called *distant bounding hyper-rectangles (DBHs)*. DBHs are the representatives of C-strings in VO. In the verification phase, the client only needs to calculate the Euclidean distance between s_q and DBHs. Since the number of DBHs is much smaller than the number of C-strings, and Euclidean distance calculation is much faster than that of edit distance, the verification cost of the $E\text{-}VS^2$ approach is much cheaper than that of VS^2. Next, we explain the details of the three phases of the $E\text{-}VS^2$ approach.

4.3.1 Verification Preparation

Before outsourcing the dataset D to the server, in addition to constructing the MB^{ed}-tree T, the data owner maps D to the Euclidean space E via a similarity-preserving embedding function $f : D \rightarrow E$. We use SparseMap Hjaltason and Samet (2003) as the embedding function due to its contractive property (Sect. 2.2). The complexity of the embedding is $O(cdn^2)$, where c is a constant value between 0 and 1, d is the number of dimensions of the Euclidean space, and n is the number of strings of D. We agree that the complexity of string embedding is comparable to the complexity of similarity search over D. However, such embedding procedure is performed only once. Its cost will be amortized over the authentication of all the future record matching verification.

The data owner sends D, T and the embedding function f to the server. The server constructs the embedded space of D by using the embedding function f. The function f will also be available to the client for result authentication.

4.3.2 VO Construction

Given the record matching request (s_q, θ) from the client, the server applies the embedding function f on s_q, and finds its corresponding node P_q in the

Euclidean space. Then the server finds the result set M^S of s_q. To prove the authenticity, soundness and completeness of M^S, the server builds a verification object VO. First, similar to VS^2, the server searches the MB^{ed}-tree to find MF-nodes of the NC-strings. For the C-strings, the server constructs a set of *distant bounding hyper-rectangles* (DBHs) from their embedded nodes in the Euclidean space. Before we define DBH, first, we define the minimum distance between an Euclidean point and a hyper-rectangle. Given a set of points $\mathcal{P} = \{P_1, \ldots, P_t\}$ in a d-dimensional Euclidean space, a hyper-rectangle $R(\langle l_1, u_1\rangle, \ldots, \langle l_d, u_d\rangle)$ is the *minimum bounding hyper-rectangle* (MBH) of $\mathcal{P}$ if $l_i = min_{k=1}^t(P_k[i])$ and $u_i = max_{k=1}^t(P_k[i])$, for $1 \leq i \leq d$, where $P_k[i]$ is the i-dimensional value of P_k. For any point P and any hyper-rectangle $R(\langle l_1, u_1\rangle, \ldots, \langle l_d, u_d\rangle)$, the minimum Euclidean distance between P and R is

$$dst^E_{min}(P, R) = \sqrt{\sum_{1\leq i\leq d} m[i]^2}, \tag{3}$$

where $m[i] = max\{l_i - p[i], 0, p[i] - u_i\}$. Intuitively, if the node P is inside R, the minimum distance between P and R is 0. Otherwise, we pick the length of the shortest path that starts from P to reach R. We have:

Lemma 2. *Given a point P and a hyper-rectangle R, for any point $P' \in R$, the Euclidean distance $dst^E(P', P) \geq dst^E_{min}(P, R)$.*

The proof of Lemma 2 is trivial. We omit the details due to its straightforwardness.

Now we are ready to define *distant bounding hyper-rectangles* (DBHs).

Definition 4. *Given a target record s_q, let P_q be its embedded point in the Euclidean space. For any hyper-rectangle R in the same space, R is a* distant bounding hyper-rectangle *(DBH) of P_q if $dst^E_{min}(P_q, R) > \theta$.*

Given a DBH R, Lemma 2 guarantees that $dst^E(P_q, P) > \theta$ for any point $P \in R$. Recalling the contractive property of the SparseMap method, we have $dst^E(P_i, P_j) \leq DST(s_i, s_j)$ for any string pair s_i, s_j and their embedded points P_i and P_j. Thus we have the following theorem:

Theorem 2. Given a target string s_q, let P_q be its embedded point. Then for any string s, s must be dissimilar to s_q if there exists a DBH R of P_q such that $P \in R$, where P is the embedded point of s.

Based on Theorem 2, to prove that the C-strings are dissimilar to the target record s_q, the server can build a number of DBHs to the embedded Euclidean points of these C-strings. We must note that not all C-strings can be included into DBHs. This is because the embedding function may introduce *false positives*, i.e., there may exist a C-string s of s_q whose embedded point P becomes similar to P_q, where P_q is the embedded point of s_q. Given a target string s_q, we can categorize a C-string s of s_q into one of the two types:

- ***E*-string:** if $dst^E(P, P_q) > \theta$, where P and P_q are the embedded points of s and s_q respectively; or

– NE**-string:** if $dst^E(P, P_q) \leq \theta$.

Therefore, given a target string s_q and a set of C-strings, first, the server classifies them into E- and NE-strings, based on the Euclidean distance between their embedded points and the embedded point of s_q. Apparently the embedded points of NE-strings cannot be put into any DBH. Therefore, the server only considers E-strings and builds DBHs, so that the client can learn that none of the E-strings are similar to s_q from the DBHs only. In order to minimize the verification cost at the client side, the server aims to minimize the number of DBHs. We formally state the problem as following.

MDBH Problem: Given a set of E-strings $\{s_1, \ldots, s_t\}$, let $\mathcal{P} = \{P_1, \ldots, P_t\}$ be their embedded points. Construct a minimum number of DBHs $\mathcal{R} = \{R_1, \ldots, R_k\}$ such that: (1) $\forall R_i, R_j \in \mathcal{R}$, R_i and R_j do not overlap; and (2) $\forall P_i \in \mathcal{P}$, there exists a DBH $R \in \mathcal{R}$ such that $P_i \in R$.

Next, we present the solution to the $MDBH$ problem. We first present the simple case for the 2-D space (i.e. $d = 2$). Then we discuss the scenario when $d > 2$. For both settings, consider the same input that includes a query point P_q and a set of Euclidean points $\mathcal{P} = \{P_1, \ldots, P_t\}$ which are the embedded points of a target string s_q and its E-strings respectively.

When $d = 2$**.** We construct a graph $G = (V, E)$ such that for each point $P_i \in \mathcal{P}$, it corresponds to a vertex $v_i \in V$. For any two vertices v_i and v_j that correspond to two points P_i and P_j, there is an edge $(v_i, v_j) \in E$ if $dst^E_{min}(P_q, R) > \theta$, where R is the MBH of P_i and P_j. We have:

Theorem 3. Given the graph $G = (V, E)$ constructed as above, for any clique C in G, let R be the MBH constructed from the points corresponding to the vertice in C. Then R must be a DBH.

Due to the space limit, the proof of Theorem 3 is omitted. But it can be found in our full paper Dong and Wang (2016). Based on Theorem 3, the *MDBH* problem is equivalent to the well-known *clique partition problem*, which is to find the smallest number of cliques in a graph such that every vertex in the graph belongs to exactly one clique. The clique partition problem is NP-complete. Thus, we design the heuristic solution to our *MDBH* problem. Our heuristic algorithm is based on the concept of *maximal cliques*. Formally, a clique is maximal if it cannot include one more adjacent vertex. The maximal cliques can be constructed in polynomial time Eidenbenz and Stamm (2000). It is shown that every maximal clique is part of some optimal clique-partition Eidenbenz and Stamm (2000). Based on this, finding a minimal number of cliques is equivalent to finding a number of maximal cliques. Thus we construct maximal cliques of G iteratively, until all the vertices belong to at least one clique.

When $d > 2$**.** Unfortunately, Theorem 3 can not be extended to the case of $d > 2$. There may exist MBHs of the pairs (v_i, v_j), (v_i, v_k), and (v_j, v_k) that are DBHs. However, the MBH of the triple (v_i, v_j, v_k) is not a DBH, as it includes a point w such that w is not inside $R(v_i, v_j)$, $R(v_i, v_k)$, and $R(v_j, v_k)$, but $dst^E(P_q, w) < \theta$.

To construct the DBHs for the case $d > 2$, we slightly modify the clique-based construction algorithm for the case $d = 2$. In particular, when we extend a clique C by adding an adjacent vertex v, we check if the MBH of the extended clique $C' = C \cup \{v\}$ is a DBH. If not, we delete the edge (u, v) from G for all $u \in C$. This step ensures that if we merge any vertex into a clique C, the MBH of the newly generated clique is still a DBH.

For both cases $d = 2$ and $d > 2$, the complexity of constructing DBHs from DBH-strings is $O(n_E^3)$, where n_E is the number of DBH-strings.

Now we are ready to describe VO construction by the E-VS^2 approach. Given a dataset D and a target string s_q, let M^S and F be the similar strings and false hits of s_q respectively. VS^2 categorizes F into C-strings and NC-strings. E-VS^2 further groups C-strings into E-strings and NE-strings. Then E-VS^2 constructs VO from M^S, NC-strings, E-strings, and NE-strings separately. Formally,

Definition 5. *Given a target string s_q, let M^S be the returned similar strings of s_q. Let $\mathcal{R}$ be the set of DBHs constructed from the E-strings. Then the VO of s_q consists of:*

(i) string s, for each M-string or NE-string;
(ii) a pair $([N_b, N_e], h^{1\to f})$ for each MF-node N, where $[N_b, N_e]$ denote the string range associated with N, and $h^{1\to f} = h(h_{C_1}||\ldots||h_{C_f})$, with $C_1, \ldots, C_f$ being the children of N;
(iii) the set of DBHs $\mathcal{R}$; and
(iv) a pair (s, p_R) for each E-string, where p_R is the pointer to the DBH in $\mathcal{R}$ that covers the Euclidean point of s.

Furthermore, in VO, a pair of square bracket is added around the strings that are located in the same node of T.

Example 4. *Consider the MB^{ed}-tree in Fig. 3 (a). The target record is s_q. The similar strings $M^S = \{s_2\}$. The only MF-node is N_7 that covers the NC-strings s_{10}, s_{11}, s_{12}. Consider the embedded Euclidean space shown in Fig. 3 (b). Apparently s_3, s_5, s_6, s_8 are the NE-strings as the Euclidean distance between them and P_q is no larger than θ. While the E-strings are s_1, s_4, s_7, s_9. The two DBHs, i.e., R_1 and R_2, constructed from these E-strings are shown as the rectangles in Fig. 3 (b). Therefore, the VO of target string s_q is*

$$\begin{aligned} VO = \{((((s_1, p_{R_1}), s_2, s_3), ((s_4, p_{R_2}), s_5, s_6)), (((s_7, p_{R_1}),\\ s_8, (s_9, p_{R_2})), ([s_{10}, s_{12}], h^{10\to 12}))), \{R_1, R_2\}\},\\ \textit{where } h^{10\to 12} = H(H(s_{10})||H(s_{11})||H(s_{12})). \end{aligned}$$

4.3.3 VO Verification

After receiving M^S and VO from the server, the client uses VO to verify if M^S is authentic, sound and complete. The verification of E-VS^2 consists of four steps. The first three steps are similar to the three steps of the VS^2 approach. The fourth step is to re-compute a set of Euclidean distance. Next, we discuss the four steps in details.

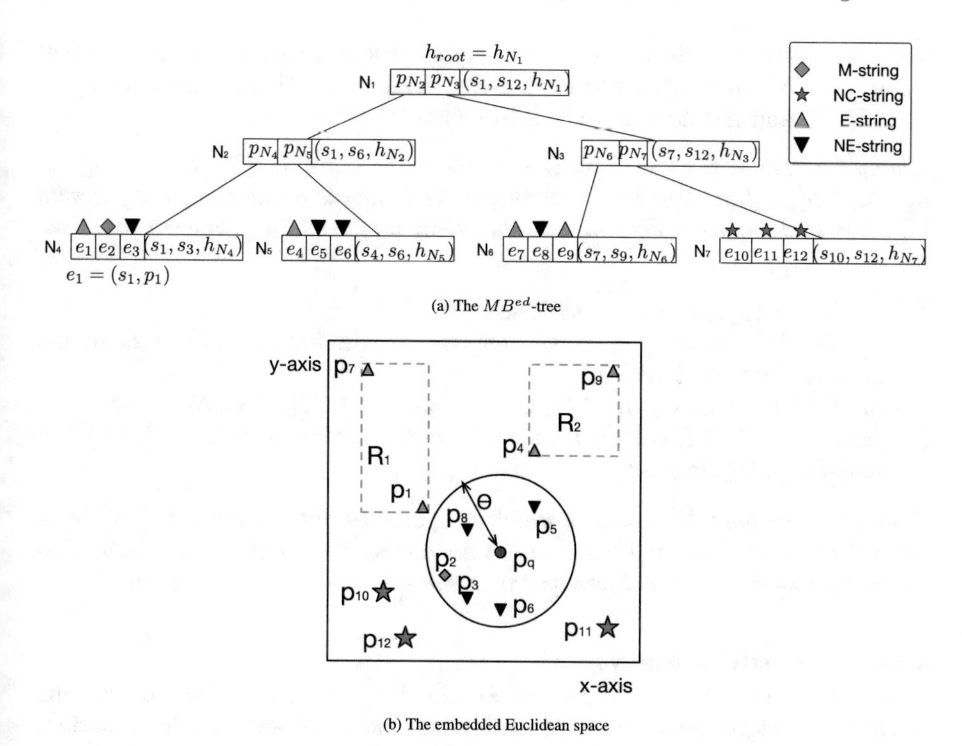

(a) The MB^{ed}-tree

(b) The embedded Euclidean space

Fig. 3. An example of VO construction by $E\text{-}VS^2$ method

Step 1 & 2: These two steps are exactly the same as Step 1 & 2 of VS^2.

Step 3: Re-computing necessary edit distance. Similar to VS^2, first, for each $s \in M^S$, the client verifies $DST(s, s_q) \leq \theta$. Second, for each range $[N_b, N_e] \in VO$, the client verifies whether $DST_{min}(s_q, N) > \theta$, where N is the corresponding MB^{ed}-tree node associated with the range $[N_b, N_e]$. The only difference of the $E\text{-}VS^2$ approach is that for each NE-string s, the client verifies if $DST(s_q, s) > \theta$. If not, the client concludes that M^S is incomplete.

Step 4: Re-computing of necessary Euclidean distance. Step 3 only verifies the dissimilarity of NE- and NC-strings. In this step, the client verifies the dissimilarity of E-strings. First, for each pair $(s, p_R) \in VO$, the client checks if $P_s \in R$, where P_s is the embedded point of s, and R is the DBH that p_R points to. If all pairs pass the verification, the client ensures that the DBHs in VO covers the embedded points of all the E-strings. Second, for each hyper-rectangle $R \in VO$, the client checks if $dst^E_{min}(P_q, R) > \theta$. If it is not, the client concludes that the returned results are not complete.

Note that we do not require the client to re-compute the edit distance between any E-string and the target string. Instead we only require the computation of the Euclidean distance between a small set of DBHs and the embedded points of the target string. Since the Euclidean distance computation is much faster than that of the edit distance (20 times more efficient according to our empirical study), $E\text{-}VS^2$ saves much verification cost compared with VS^2. More comparison of VS^2 and $E\text{-}VS^2$ can be found in Sect. 5.

Example 5. Following the running example in Example 4, after calculating the root hash h'_{root} from VO in the same way as Example 3 and comparing it with h_{root} received from the data owner, the client performs the following computations:

(1) for $M^S = \{s_2\}$, compute $DST(s_q, s_2)$;
(2) for the NE-strings s_3, s_5, s_6, s_8, compute the edit distance between them and s_q, and compare it with θ;
(3) for the pair $([s_{10}, s_{12}], h^{10\to 12}) \in VO$, compute $DST_{min}(s_q, N_7)$; and
(4) for the two DBH in VO, namely R_1, and R_2, check if $dst^E_{min}(P_q, R_1) > \theta$ and $dst^E_{min}(P_q, R_2) > \theta$.

Compared with the VS^2 approach which computes 10 edit distances, $E\text{-}VS^2$ computes 6 edit distances, and 2 Euclidean distances. Note that the Euclidean distances computation is much cheaper than the edit distance computation.

4.3.4 Robustness Analysis

Similar to the robustness discussion for the VS^2 approach (Sect. 4.2.4), the server may perform three types of cheating behaviors, i.e., authenticity violation, soundness violation and completeness violation. Next, we prove that $E\text{-}VS^2$ can catch any of the cheating behaviors.

Authenticity. The authenticity verification is similar to that of VS^2. Any authenticity violation can be detected in Step 2 of the verification procedure, as the re-constructed root value from the tampered values must not match the original one.

Soundness. Similar to VS^2, any soundness violation can be caught in Step 3 where the client calculates the edit distance for every string in M^S.

Completeness. In addition to the three cases of completeness violation in VS^2, in $E\text{-}VS^2$, we have Case 4: the server returns the incomplete result $M^S = M - \{s'\}$ for some string $s' \in D$ and $s' \approx s_q$, and claims that s' is an E-string by including s' into a DBH R. Next, we have the following theorem to demonstrate the robustness of $E\text{-}VS^2$.

Theorem 4. Given a target record s_q and a DBH R (i.e., $dst^E_{min}(P_q, R) > \theta$), including P' into R will change it to a non-DBH, where P' is the embedded point of a string s' such that $s' \in D$ and $s' \approx s_q$.

Proof. According to the contractive property of the embedding function, we have $dst^E(P', P_q) \leq DST(s', s_q) \leq \theta$. Lemma 2 demonstrates that for any MBH R, it must be true that $dst^E_{min}(P_q, R) \leq dst^E(P_q, P)$ for any point $P \in R$. Therefore, by including P' into R, we have $dst^E_{min}(P_q, R) \leq dst^E(P_q, P') \leq DST(s', s_q) \leq \theta$. Hence, R is not a DBH any more. ∎

Following Theorem 4, we conclude that the client can easily catch the incomplete violation by re-computing the euclidean distance between P_q and every hyper-rectangle $R \in VO$ (Step 4 of verification).

4.3.5 Security Analysis

If the adversary $\mathcal{A}$ breaks the security of the $E\text{-}VS^2$ authentication approach by generating the incorrect matching result M' and counterfeit VO π' that pass the verification process, it must be true that all the DBHs in π' are dissimilar to P_q in the Euclidean space, and $h_{root} = h'_{root}$, where P_q is the embedded point of the target string s_q, and h'_{root} is the root digest of the MB^{ed}-tree re-constructed from π'. The security proof of $E\text{-}VS^2$ is similar to that of VS^2 (Sect. 4.2.5). Each successful break of $\mathcal{A}$ yields at least a collision of the hash function $H(\cdot)$. Thus, we can prove the security of $E\text{-}VS^2$ under the assumption of collision resistant hash functions. We omit the proof due to the space limit.

4.4 Gram Vector Authentication: $G\text{-}VS^2$

Although the $E\text{-}VS^2$ approach dramatically reduces the verification complexity at the client side by substituting the edit distance calculation of a large number of E-strings with a small amount of Euclidean distance computation for the DBHs, the number of NE-strings still can be large, and thus lead to high verification cost. Figure 4 shows the number of NE-strings with regard to the similarity threshold θ on two real-world dataset that include $260K$ and 1 million records respectively. We observe that the amount of NE-strings rises sharply with the growth of θ. On both datasets, when $\theta \geq 5$, more than 90% of the strings in the dataset are NE-strings. Considering that in $E\text{-}VS^2$, the client has to calculate the edit distance between the target string and every NE-string, the overhead can be significant.

In order to reduce the verification cost incurred by the large amount of NE-strings, we design a new approach named $G\text{-}VS^2$. The key idea of $G\text{-}VS^2$ is that, in the VO verification phase, we substitute the edit distance calculation for a large fraction of NE-strings with efficient q-gram distance computation. Our empirical evaluation shows that the q-gram distance computation is 150 times

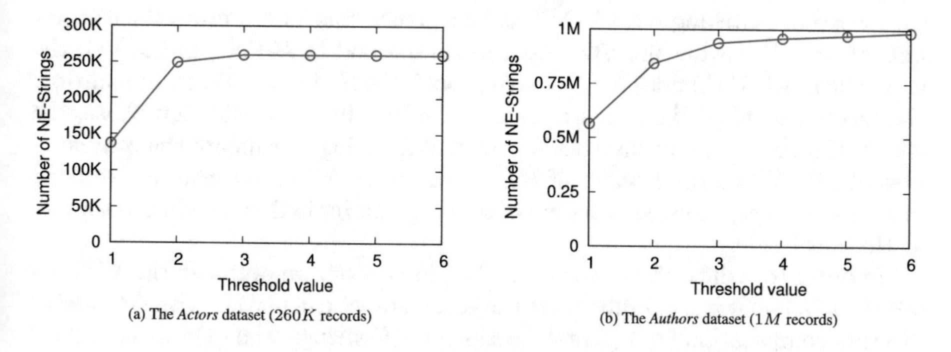

(a) The *Actors* dataset ($260K$ records)

(b) The *Authors* dataset ($1M$ records)

Fig. 4. Number of NE-strings w.r.t. similarity threshold θ

more efficient than edit distance calculation. Therefore, compared with $E\text{-}VS^2$, the $G\text{-}VS^2$ approach can substantially reduce the verification complexity at the client side.

Next, we discuss the details of $G\text{-}VS^2$ in three phases of the authentication process.

4.4.1 Verification Preparation

Before outsourcing the database D to the server, besides constructing the MB^{ed}-tree T and building the embedding function f, the data owner constructs the *gram vector* and *RSA signature* of every string in D. For any string $s_i \in D$, let $Q(s_i)$ be the *gram vector* of s_i (Sect. 2.3). Obviously, $|Q(s_i)| \leq |s_i|+q-1$, where q is the constant for q-grams. It will lead to linear space overhead if we simply store the q-grams. In order to alleviate the storage overhead, we transform the linear-size *gram vector* into the constant-size *gram counting vector* (GC-vector) based on hash function. Specifically, given a constant L, the data owner generates a hash function H_1 that maps every q-gram into a integer in the range of $[1, L]$. Formally, the GC-vector v_i of a string s_i is a vector of length L such that $v_i[k] = \mid \{1 \leq j \leq |s_i|+q-1 \mid H_1(Q(s_i)[j]) = k\} \mid$ for $1 \leq k \leq L$. In other words, $v_i[k]$ records the number of q-grams of s_i that are placed in the k-th bucket by the hash function H_1.

Next, the data owner generates a pair of private and public keys (sk, pk) of RSA. For each string s_i and its GC-vector v_i, the data owner computes its signature as

$$\sigma_i = \left(H_2(s_i||v_i)\right)^d \ (mod\ N), \tag{4}$$

where H_2 is a full domain hash function $\{0,1\}^* \rightarrow \mathcal{Z}_N$, and d and N are part of the RSA keys.

Finally, the data owner transmits the dataset D, the MB^{ed}-tree T, the embedding function f and the signatures $\{\sigma_i\}$ to the server. Besides, the data owner publishes the MB^{ed}-tree's root hash value h_{root}, f and the RSA public key pk.

4.4.2 VO Construction

Upon receiving a record matching request (s_q, θ) from a client, the server searches for the set of matching records M^S and constructs the VO to prove the correctness of M^S. Following the VO construction method in $E\text{-}VS^2$, in the VO, the server includes MF-nodes for NC-strings and DBHs for E-strings. In addition, the server inserts all the NE-strings into the VO. In the verification phase, the client calculates the edit distance for every NE-string to validate the completeness of M^S. When the number of NE-strings is large, the verification overhead may not be acceptable, especially considering the limited computing resources at the client side.

In order to reduce the verification cost, in $G\text{-}VS^2$, we optimize the VO construction in two ways: (1) substitute the edit distance calculation with GC-vector distance computation for a (large) fraction of NE-strings; and (2) further reduce the number of GC-vector distance calculation by providing a small number of representatives. Before we discuss the VO construction procedure, we first define two distance metrics based on gram vectors.

Given two strings s_1 and s_2, let Q_1 and Q_2 be the q-grams. We define the q-gram distance between Q_1 and Q_2 as

$$dst^G(Q_1, Q_2) = \frac{max(|s_1|, |s_2|) - 1 - |Q_1 \cap Q_2|}{q} + 1, \quad (5)$$

where q is the constant for q-grams, and $|Q_1 \cap Q_2|$ is the cardinality of the intersection between Q_1 and Q_2. According to Lemma 1, we have the following corollary.

Corollary 1. *For any pair of strings s_1 and s_2, it must be true that $DST(s_1, s_2) \geq dst^G(Q_1, Q_2)$.*

Considering the GC-vector construction method, we define the distance of two GC-vectors v_1 and v_2 as

$$dst^{GC}(v_1, v_2) = \frac{max(U_1(v_1, v_2), U_2(v_1, v_2))}{q}, \quad (6)$$

where L is the length of the GC-vectors,
$U_1(v_1, v_2) = \sum_{1 \leq i \leq L, v_1[i] \geq v_2[i]} v_1[i] - v_2[j]$, and $U_2(v_1, v_2) = \sum_{1 \leq i \leq L, v_1[i] < v_2[i]} v_2[i] - v_1[j]$. We have Lemma 3 to describe the relationship between edit distance and GC-vector distance.

Lemma 3. *For any pair of strings s_1 and s_2, it must be true that $DST(s_1, s_2) \geq dst^G(Q_1, Q_2) \geq dst^{GC}(v_1, v_2)$.*

Proof. Because of Corollary 1, we only need to prove $dst^G(Q_1, Q_2) \geq dst^{GC}(v_1, v_2)$. Considering the gram counting vector construction procedure, it is easy to see that $\sum_{j=1}^{L} v_1[j] = |Q_1| = |s_1| + q - 1$. It is the same for s_2. Due to the potential collision introduced by the hash function H_1, it must be true that $\sum_{j=1}^{L} min(v_1[j], v_2[j]) \geq |Q_1 \cap Q_2|$. Combining them together, we have the following inference.

$$\begin{aligned}
dst^G(Q_1, Q_2) &= \frac{max(|s_1|, |s_2|) - 1 - |Q_1 \cap Q_2|}{q} + 1 \\
&\geq \frac{max(\sum_{j=1}^{L} v_1[j] - \sum_{j=1}^{L} min(v_1[j], v_2[j]), \sum_{j=1}^{L} v_2[j] - \sum_{j=1}^{L} min(v_1[j], v_2[j]))}{q} \\
&= \frac{max(U_1(v_1, v_2), U_2(v_1, v_2))}{q} \\
&= dst^{GC}(v_1, v_2).
\end{aligned}$$

∎

Based on Lemma 3, for any NE-string s, we can categorize it into one of the two new types.

- G-**string:** if s is dissimilar to the target according to the GC-vector distance, i.e., $dst^{GC}(v, v_q) > \theta$, where v and v_q are the GC-vectors of s and s_q respectively; or
- FP-**string:** if $dst^{GC}(v, v_q) \leq \theta$.

It is straightforward that for any G-string s_i, the server can prove $s_i \not\approx s_q$ based on its GC-vector v_i. By simply calculating $dst^{GC}(v_i, v_q)$, the client is assured that s_i should not be included in M^S. Even though it is computationally efficient to compute $dst^{GC}(.)$, considering the potentially large amount of G-strings, it would be tedious for the client to compute the GC-vector distance for all the G-strings. As a remedy, we design a method for the server to organize the large number of G-strings into a small amount of *GC-groups* (in short for *gram counting vector groups*), so that in the verification process, the client only needs to calculate the GC-vector distance for the GC-groups. Next, we present the GC-groups construction process.

Let $V = \{v_1, \ldots, v_k\}$ be a sequence of GC-vectors. We create two L-dimensional bounding vectors lb and ub for V, such that $lb[j] = min_{i=1}^{k} v_i[j]$ and $ub[j] = max_{i=1}^{k} v_i[j]$ for $1 \leq j \leq L$. Given a target string s_q, let v_q be its GC-vector. Then the minimum distance between v_q and any vector $v_i \in V$ is

$$dst_{min}^{GC}(v_q, V) = \frac{max(W_1(v_q, lb), W_2(v_q, ub))}{q}, \tag{7}$$

where $W_1(v_q, lb) = \sum_{1 \leq j \leq L,\ v_q[j] < lb[j]} lb[j] - v_q[j]$, and $W_2(v_q, ub) = \sum_{1 \leq j \leq L,\ v_q[j] > ub[j]} v_q[j] - ub[j]$. Clearly, for any $v_i \in V$, it must be true that $dst^{GC}(v_q, v_i) \geq dst_{min}^{GC}(v_q, V)$. For any GC-vector sequence V, we call V a *distant bounding group* (DBG) if $dst_{min}^{GC}(v_q, V) > \theta$.

In order to minimize the number of GC-vector distance calculation at the client side, we aim at finding an efficient approach to create a small number of DBGs from the set of GC-vectors $\{v_i\}$ of the G-strings. Next, we propose the approach that is based on space-filling curve and dynamic programming.

We first use a locality-preserving space-filling curve (e.g., the Z-order curve Morton (1966) or Hilbert curve Kamel and Faloutsos (1993); Moon et al. (2001)) on the set of GC-vectors to generate the one-dimensional indices of the L-dimensional vectors. Let $\varphi(.)$ be the indexing function from the indexing curve. The locality-preserving property of the curves guarantee that similar GC-vectors have close indices.

Given a set of G-strings $\{s_1, \ldots, s_t\}$, let $W = \{v_1, \ldots, v_t\}$ be their gram counting vectors arranged according to the indices on the space-filling curve, our objective is to partition W into the smallest number of DBGs. In other words, we aim at finding $\mathcal{V} = \{V_1, \ldots, V_z\}$, where (1) $V_1 \cup \cdots \cup V_z = W$; (2) $V_i \cap V_j = \emptyset$ for any $i \neq j$; (3) for any $v_j \in V_i$, and $v_k \in V_{i+1}$, let j and k be their curve indices respectively, it must hold that $j < k$; and (4) z is minimum. Obviously, this problem exhibits the *optimal substructure property*

Cormen (2009), i.e., the optimal solutions to a problem must incorporate the optimal solutions to related subproblems, which can be resolved independently. Therefore, we design a dynamic programming algorithm to solve the problem.

We use $W[i,\ldots,j] = \{v_i,\ldots,v_j\}$ to denote the subsequence of the gram counting vectors. Let c and e be two two-dimensional arrays, where $c[i,j]$ denotes the minimum number of DBGs for $W[i,\ldots,j]$, whereas $e[i,j]$ denotes the index of the first *GC*-vector in the last DBG of the optimal solution for $W[i,\ldots,j]$. Besides, we use $V_{i,j}$ to represent the single bounding group directly from $\{v_i,\ldots,v_j\}$. We present our algorithm named *DP-DBG* in Algorithm 1 to construct the minimum number of DBGs for W.

In line 2–4 of Algorithm 1, we initialize c as an identity matrix, and $e[i,i] = i$. This is because we only need a single DBG for a single *GC*-vector. Starting from line 5, we iterative increase ℓ, which is the number of elements in W that the DBGs can cover. In line 8, if after inserting v_j into the last DBG of the optimal solution for $W[i,\ldots,j-1]$, the new bounding group $V_{e[i,j-1],j}$ is dissimilar to v_q, the minimum number of DBGs for $W[i,\ldots,j]$ is the same as that for $W[i,\ldots,j-1]$. Otherwise, the optimal solution for $W[i,\ldots,j]$ would be a combination of the optimal solutions for $W[i,\ldots,k]$ and $W[k+1,\ldots,j]$ for some $i \leq k < j$,

---

Algorithm 1. *DP-DBG*(v_q, W)

Require: The gram counting vector v_q of the target string s_q, the sequence of gram counting vectors $W = \{v_1,\ldots,v_t\}$ ordered by curve indices
Ensure: The minimum number of DBGs $\mathcal{V}$ for W

1: Let $c[1\ldots t, 1\ldots t]$ and $e[1\ldots t, 1\ldots t]$ be new tables
2: **for** $i = 1$ to t **do**
3: $\quad c[i,i] = 1$
4: $\quad e[i,i] = i$
5: **for** $\ell = 1$ to $t-1$ **do**
6: $\quad$ **for** $i = 1$ to $t-\ell+1$ **do**
7: $\quad\quad j = i + \ell$
8: $\quad\quad$ **if** $dst^{GC}_{min}(v_q, V_{e[i,j-1],j}) > \theta$ **then**
9: $\quad\quad\quad c[i,j] = c[i,j-1]$
10: $\quad\quad\quad e[i,j] = e[i,j-1]$
11: $\quad\quad$ **else**
12: $\quad\quad\quad c[i,j] = \infty$
13: $\quad\quad\quad e[i,j] = j+1$
14: $\quad\quad\quad$ **for** $k = i$ to $j-1$ **do**
15: $\quad\quad\quad\quad$ **if** $c[i,k] + c[k+1,j] < c[i,j]$ **then**
16: $\quad\quad\quad\quad\quad c[i,j] = c[i,k] + c[k+1,j]$
17: $\quad\quad\quad\quad\quad e[i,j] = k+1$
18: $\mathcal{V} = \emptyset$
19: $i = t$
20: **while** $i > 0$ **do**
21: $\quad \mathcal{V} = \mathcal{V} \cup \{V_{e[1,i],i}\}$
22: $\quad i = e[1,i] - 1$
23: **return** $\mathcal{V}$

---

whichever results in the minimum number of DBGs (line 12–17). Starting from line 18, we create the optimal solution $\mathcal{V}$ for W from the table e.

The DBGs $\mathcal{V}$ alone are not sufficient to demonstrate that all G-strings are dissimilar to s_q, as it is possible that the DBGs in the VO are not constructed from the G-strings. In order for the server to prove that, one straightforward approach is that the server includes a triplet (s_i, v_i, σ_i) for every G-string in the VO, where σ_i is the RSA signature computed by the data owner according to Eq. (4). The client verifies the authenticity of s_i and v_i by checking if $(\sigma_i)^e \stackrel{?}{=} H_2(s_i||v_i)\ (mod\ N)$, where e and N are the RSA public key. Also, the client checks if v_i belongs to any $V_j \in \mathcal{V}$. However, this approach introduces considerable network bandwidth and verification overhead. Let n_G denote the number of G-strings. The server delivers $O(n_G)$ RSA signatures to the client. Following that, the client executes $O(n_G)$ modular exponentiations to verify all pairs of G-strings and GC-vectors.

To reduce the VO size and verification cost, we adapt the *Condensed-RSA* (Sect. 2.5). For all the G-strings, the server only constructs a single aggregate signature:

$$\sigma_G = \prod_{i=1}^{n_G} \sigma_i\ (mod\ N). \tag{8}$$

The server substitutes $\{\sigma_i\}$ with σ_G in VO. Now we are ready to formally define the VO.

Definition 6. *Given a target record s_q, let M^S be the returned similar strings of s_q, $\mathcal{R}$ be the set of DBHs constructed from the E-strings, and $\mathcal{V}$ be the set of DBGs constructed from the G-strings. The VO of M^S consists of:*

(i) string s, for each M-string or FP-string;
(ii) a pair $([N_b, N_e], h^{1\to f})$ for each MF-node N, where $[N_b, N_e]$ denote the string range associated with N, and $h^{1\to f} = h(h_{C_1}||\ldots||h_{C_f})$, with $C_1, \ldots, C_f$ being the children of N;
(iii) the set of DBRs $\mathcal{R}$ and DBGs $\mathcal{V}$;
(iv) a pair (s, p_R) for each E-string s, where p_R is the pointer to the DBH $R \in \mathcal{R}$ that covers the Euclidean point of s;
(v) a triplet (s, v, p_V) for each G-string, where v is the GC-vector of s, and p_V is the pointer to the DBG $V \in \mathcal{V}$ that covers v;
(vi) the aggregate signature σ_G for all the G-strings.

Furthermore, in VO, a pair of square bracket is added around the strings that are located in the same node of T.

Example 6. Consider the MB^{ed}-tree and GC-vectors in Fig. 5 (a) and (b). The G-strings are s_3, s_6 and s_8, while the only FP-string is s_5. Following the DBG construction procedure in Algorithm 1, the server generates a single DBG V_1 to

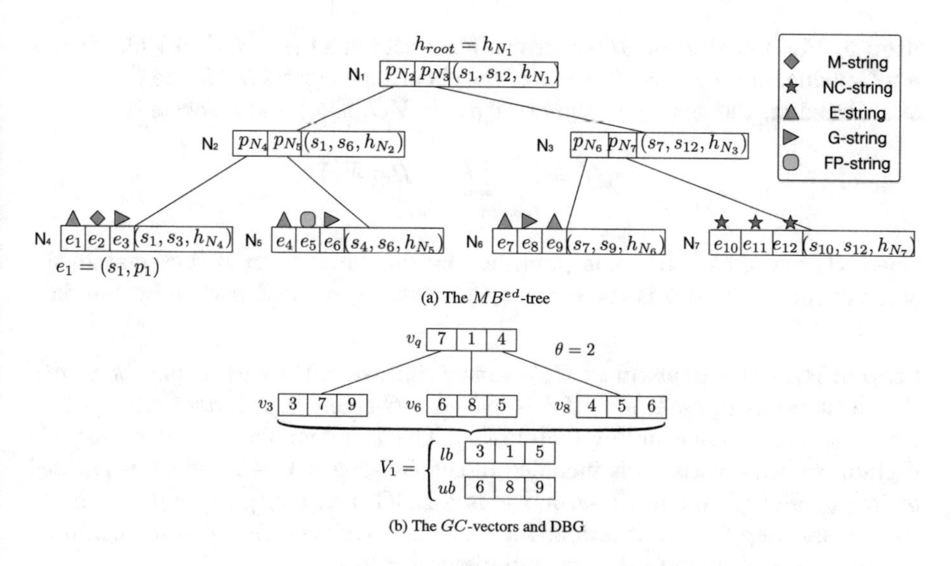

(a) The MB^{ed}-tree

(b) The GC-vectors and DBG

Fig. 5. An example of VO construction by $G\text{-}VS^2$ method

cover all the G-strings. In addition, the server computes the aggregated signature $\sigma_G = \sigma_3 \times \sigma_6 \times \sigma_8$ *(mod N). Therefore, the VO of the target string* s_q *is*

$$\begin{aligned}VO = \{((((s_1, p_{R_1}), s_2, (s_3, v_3, p_{V_1})), ((s_4, p_{R_2}), s_5, (s_6, v_6,\\ p_{V_1}))), (((s_7, p_{R_1}), (s_8, v_8, p_{V_1}), (s_9, p_{R_2})), ([s_{10}, s_{12}],\\ h^{10\to12}))), \{R_1, R_2\}, \{V_1\}, \sigma_G\},\\ \text{where } h^{10\to12} = H(H(s_{10})||H(s_{11})||H(s_{12})).\end{aligned}$$

4.4.3 VO Verification

When the client receives the matching result M^S and VO from the server, the client checks the authenticity, soundness and completeness of M^S based on VO. The verification procedure consists of six steps, i.e., Step 1 reconstruction of MB^{ed}-Tree, Step 2 re-computation of root hash, Step 3 re-computation of necessary edit distance, Step 4 re-computation of necessary Euclidean distance, Step 5 verification of GC-vectors, and Step 6 re-computation of GC-vector distance. Step 1, 2 and 4 are exactly the same as those of the $E\text{-}VS^2$ approach (Sect. 4.3). Next, we only focus on the three steps that are different from $E\text{-}VS^2$.

Step 3: Re-computation of necessary edit distance. First, the soundness verification relies on the re-computation of edit distance. In specific, for each $s \in M^S$, the client calculates $DST(s, s_q)$ and checks if it is no larger than the threshold θ. After that, for each MF-node N in VO, the client calculates $DST_{min}(s_q, N)$ to inspect if all strings in the subtree of N are dissimilar to s_q. At last, for each FP-string s that is dissimilar to s_q, but is not either a E-string or G-string, the client has to calculate $DST(s, s_q)$ to be assured that $s \not\approx s_q$.

Step 5: Verification of *GC*-vectors. For each triplet (s, v, p_V) in VO, where s is a G-string, and v is the GC-vector, the client investigates if v is the GC-vector of s. Based on the aggregate signature σ_G in VO, the client inspects if

$$(\sigma_G)^e \stackrel{?}{=} \prod_{(s,v,p_V)\in VO} H_2(s||v), \tag{9}$$

where H_2 is the hash function published by the data owner. If they match, the client is certain that v is the correct GC-vector generated from s by the data owner.

Step 6: Re-computation of *GC*-vector distance. For each triplet (s, v, p_V), the client needs to make sure if s is indeed a G-string, i.e., if $dst^{GC}(v, v_q) > \theta$, where v_q is the gram counting vector of s_q. This is accomplished by checking: (1) if gram counting vector v is included in the GC-group $V \in \mathcal{V}$, which is pointed to by p_V; and (2) if the GC-group V is a DBG, i.e., $dis^{GC}_{min}(v_q, V) > \theta$. If the VO passes Step 6 of verification, the client is convinced that the strings in all triplets of the VO are G-strings, and dissimilar to s_q.

Compared with the verification procedure of E-VS^2, to verify completeness, in G-VS^2 we do not require the client to calculate the edit distance for all NE-strings. Instead, the client only needs to calculate the edit distance for FP-strings, which is a small subset of the NE-strings. The rest of the NE-strings, which are those G-strings, are verified by calculating the GC-vector distance between v_q and a limited amount of DBGs. Our empirical study demonstrates that a small amount of DBGs can cover plenty of G-strings. For instance, we only need 5,604 DBGs for 420,673 G-strings. In other words, based on the DBG construction method, we can reduce the number of gram counting vector distance calculations at the client side by a magnitude of 2. Besides, considering the computation of gram counting vector distance is 150 times more efficient than string edit distance calculation, G-VS^2 can substantially reduce the verification cost at the client side, especially when there are numerous G-strings. More detailed complexity analysis can be found in Sect. 5.

Example 7. Following the running example in Example 6, after calculating the root hash h'_{root} from VO and comparing it against h_{root}, which is issued by the data owner, the client performs the following computations:

(1) for $M^S = \{s_2\}$, compute $DST(s_q, s_2)$;
(2) for the single FP-string s_5, compute the edit distance $DST(s_5, s_q)$, and compare it with θ;
(3) for the pair $([s_{10}, s_{12}], h^{10\rightarrow 12}) \in VO$, compute $DST_{min}(s_q, N_7)$;
(4) for the two DBH in VO, namely R_1, and R_2, check if $dst^E_{min}(P_q, R_1) > \theta$ and $dst^E_{min}(P_q, R_2) > \theta$;
(5) for the three G-strings in VO, namely s_3, s_6 and s_8, check if $(\sigma_G)^e = H(s_3||v_3) \times H(s_6||v_6) \times H(s_8||v_8) \pmod N$, where σ_G is the aggregate signature included in VO; and
(6) for the DBG V_1, check if $dst^{GC}_{min}(v_q, V_1) > \theta$.

Compared with the E-VS^2 approach which computes 6 edit distances and 2 Euclidean distances, G-VS^2 computes 3 edit distances, 2 Euclidean distances, 1 GC-vector distance and 1 modular exponentiation. Note that the edit distances computation is much more expensive than the other routines.

4.4.4 Robustness Analysis

In this section, we formally prove that with *G*-VS^2, the client is capable of catching any incorrect result M^S by verifying the VO. The authenticity and soundness analysis is the same as that of the *E*-VS^2 approach. Thus, we only focus on the completeness analysis.

Completeness. The server may exclude an M-string $s \in D$, $s \approx s_q$ from M^S. It is possible that the server excludes s from the VO, or takes s as an NC-string, or an E-string, or a G-string, or FP-string. Compared with *E*-VS^2, there is only new case, which is Case 5: the server returns the incomplete result $M^S = M - \{s'\}$ for some string $s' \in D$ and $s' \approx s_q$, and claims that s' is a G-string by including v' into a DBG, where v' is the GC-vector of s'. Next, we have the following theorem to support our robustness analysis.

If s' is claimed as a G-string by including a triplet (s', v', p_V) in VO, the server may provide the incorrect GC-vector $\hat{v}$ of s such that $dst^{GC}(v_q, \hat{v}) > \theta$. The client catches s in Step 5. Without the knowledge of the secret key of RSA scheme, the server cannot generate a condensed signature σ_G for Eq. (9) to hold. Otherwise, if the server provides the correct GC-vector v' of s', and adds v' to any DBG V, we have the following theorem.

Theorem 5. Given a target record s_q and a DBG V such that $dst^{GC}_{min}(v_q, V) > \theta$, where v_q is the GC-vector of s_q, if we add the GC-vector v' into V for any M-string s' (i.e., $DST(s_q, s') \leq \theta$), V must not be a DBG.

Proof. The proof is straightforward. For any string s whose GC-vector v is embraced by the GC-group V, according to Lemma 3 and Eq. (7), we have $dst^{GC}_{min}(v_q, V) \leq dst^{GC}(v_q, v) \leq DST(s_q, s)$. If $DST(s_q, s') \leq \theta$, it must be true that $dst^{GC}_{min}(v_q, V) \leq \theta$. Therefore, V is not a DBG. ∎

Theorem 5 guarantees that the client can catch the server's incomplete result in Step 6 after getting $dst^{GC}_{min}(v_q, V) \leq \theta$ for at least a GC-group $V \in \mathcal{V}$.

4.4.5 Security Analysis

The security analysis of the *G*-VS^2 approach is more complicated than that of the VS^2 and *E*-VS^2 approaches. The reason is mainly because the server additionally includes the aggregate RSA signature of the G-strings in the VO. In this section, we demonstrate *G*-VS^2 is secure based on the collision resistant hash functions and the intractability of the integer factorization problem.

Outline of Security. Suppose that the adversary $\mathcal{A}$ successfully generates the incorrect result M' and counterfeit VO π' that pass the verification process (Sect. 4.4.3). The following two conditions must hold: (1) the root hash value re-computed from π' matches the local copy, i.e., $h'_{root} = h_{root}$; and (2) the GC-vectors match the aggregate signature in π' (Eq. (9)). Obviously, the first condition contradicts the cryptographic assumption of collision resistant hash functions. While the second condition contradicts the security of Condensed-RSA, which has been proved based on the intractability of the integer factorization problem Mykletun et al. (2006). Therefore, we prove the security of the G-VS^2 approach.

4.5 Probabilistic Authentication: P-VS^2

The G-VS^2 approach alleviates the verification cost at the client side by avoiding the edit distance calculation for G-strings. As a result, the client only needs to calculate the edit distance for the FP-strings. Hence, in G-VS^2, the verification performance highly depends on the number of FP-strings. In Fig. 6, we display the number of FP-strings on two real-world datasets. We observe that with the growth of similarity threshold θ, the number of FP-strings increases intensively. This is because of the sharp drop in the number of G-strings. For instance, when $\theta = 6$, only 1% of the dataset are G-strings. Therefore, we can expect considerable verification cost at the client side.

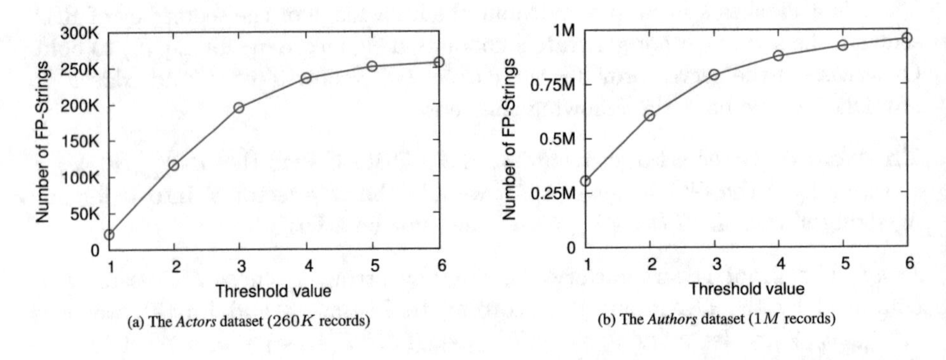

Fig. 6. Number of FP-strings w.r.t. similarity threshold θ

To mitigate the problem raised by a large number of FP-strings, it is desirable if the client only needs to calculate the edit distance for a small fraction of them for completeness verification. Motivated by this, we design a new approach named P-VS^2 that puts this aspiration into effect by building a strategic game between the server and the client. Next we formally define the game.

Definition 7. *The strategic game consists of two players: the client and the server. The server has two potential actions:* cheat *or* do not cheat *on the FP-strings.*

- **Cheat:** *By cheating on FP-strings, the server may violate result completeness by claiming at least one string $s \in D$ and $s \approx s_q$ as a FP-string.*

- **Not Cheat:** *The server honestly calculates the edit distance for all FP-strings.*

The client also has two actions: verify *or* not verify *the FP-strings.*

- **Verify:** *The client only* verify a small fraction *of the FP-strings by calculating the edit distance. We call such strings as V-strings (in short for* verified-*strings), and the rest FP-strings as NV-strings.*
- **Not Verify:** *Without any verification, the client simply believes that all the FP-strings in VO are dissimilar to the target record with a certain confidence.*

It is desired if by deliberately designing the game, there is a Nash equilibrium (Sect. 2.7) where the server chooses not to cheat on FP-strings. In this way, any rational server would always behave honestly in order to receive the maximum payoff Morris (2012). We adopt game theory because it assumes that the players are self-interested. This is consistent with our problem, in which the server and the client aim to maximizing their payoffs separately, which leads to their conflicting interests. This is different from the classic multi-agent system Parsons and Wooldridge (2002); Raimondi et al. (2005), where the players/agents are benevolent, i.e., they share a common or compatible objective, and have no conflict of interest.

Next, we discuss the payoff of each action in the strategic game. For the client, the utility of the FP-strings in VO is U. If the client trusts the FP-strings (either pass verification or accept without verification), the client pays P to the server. However, if the client catches the server's cheating behavior on any FP-string, the server needs to compensate P_T to the client as the penalty of cheating. Let n_{FP} denote the number of FP-strings in VO. The client may choose to verify the FP-strings by calculating the edit distance for α fraction ($\alpha \in (0, 1]$) of them. The cost of edit distance calculation for all FP-strings at the client and server side are C_C and C_S respectively. In the outsourcing model, typically, we have $P_T >> U >> P > C_S >> C_C$. If the client does not verify the FP-strings, the client simply puts η trust on them. Or the client may put η confidence on the FP-strings without verification at all. In practice, η can be set as the credibility of the server according to the historical performance. Table 1 summarizes the notations in the game.

Let p^{Client} and p^{Server} be the payoffs for the client and server respectively. We present (p^{Client}, p^{Server}) under different circumstances in Table 2. Next, we explain the payoffs in detail.

- **(Verify, Not Cheat):** When the server does not cheat on any FP-string in VO, certainly, the VO passes the verification. The client obtains the utility U from R^S at the expense of P (the payment to the server) and αC_C (the cost to calculate the edit distance for α fraction of FP-strings). Therefore, we have $p^{Client}_{(Verify, Not\ Cheat)} = U - P - \alpha C_C$. The server's payoff is $p^{Server}_{(Verify, Not\ Cheat)} = P - C_S$, where C_S is the edit distance computation cost for FP-strings.
- **(Verify, Cheat):** If the server cheats on FP-strings, by calculating the edit distance for α fraction of them, the client has a certain probability to detect

Table 1. Notations in the strategic game

Notation	Meaning
U	The utility of FP-strings in VO for client
P	The payment to the server if the client accepts all FP-strings
P_T	The penalty if the client catches the cheating behavior
C_S	The cost to calculate edit distance at server side
C_C	The cost to calculate edit distance at client side
n_{FP}	The number of FP-strings in VO
α	The fraction of FP-strings in VO that are verified by the client
β	The fraction of FP-strings in VO that are dissimilar to s_q
η	The confidence that the server does not cheat on FP-strings

Table 2. Payoff matrix for the client and server

		Not cheat	Cheat
Verify	Catch	N/A	$(P_T - \alpha C_C, -P_T - \beta C_S)$
	Not Catch	$(U - P - \alpha C_C, P - C_S)$	$(\beta U - P - \alpha C_C, P - \beta C_S)$
Not Verify		$(\eta U - P, P - C_S)$	$(\beta U - P, P - \beta C_S)$

the cheating behavior. We denote the probability with ϵ. The inference of ϵ will be discussed later. If the client catches it, the server receives no payment, and compensates P_T to the client according to the protocol. Let β be the fraction of FP-strings that the server conducts the edit distance calculation, then the payoff of the client and server are $P_T - \alpha C_C$ and $-P_T - \beta C_S$ respectively. Chances are that the client cannot catch the server's cheating behavior with limited fraction of edit distance calculations. In this case, the client's payoff is $\beta U - P - \alpha C_C$, because only β fraction of the FP-strings are indeed dissimilar to s_q. The server's payoff is $P - \beta C_S$.

- **(Not Verify, Not Cheat):** If the client does not check the correctness of FP-strings, the client simply accepts them with η confidence. Hence, the client's payoff is $p^{Client}_{(Not\ Verify, Not\ Cheat)} = \eta U - P$, while the server's payoff is $p^{Server}_{(Not\ Verify, Not\ Cheat)} = P - C_S$.
- **(Not Verify, Cheat):** If the server cheats on $(1 - \beta)$ fraction of the FP-strings, the client only receives β fraction of the total utility. Thus, the client's payoff is $p^{Client}_{(Not\ Verify, Cheat)} = \beta U - P$. The server's payoff is $p^{Server}_{(Not\ Verify, Cheat)} = P - \beta C_S$.

Recall that our objective is to have the action profile $(*, Not\ Cheat)$ be the equilibrium in the game, where $*$ denotes any possible action of the client. According to Table 2, $(Verify, Not\ Cheat)$ can be the only feasible equilibrium, as if the client does not verify the FP-strings, the server can gain more payoff by simply switching from *Not Cheat* to *Cheat*.

In order for the action profile $(Verify, Not\ Cheat)$ to be the equilibrium, there are two conditions to be satisfied.

- **Condition 1:** the client does not gain more payoff by switching to the action *Not Verify*, i.e.,

$$p^{Client}_{(Verify,Not\ Cheat)} \geq p^{Client}_{(Not\ Verify,Not\ Cheat)}. \tag{10}$$

To let Condition 1 hold, according to Table 2, we must have

$$\alpha \leq \frac{(1-\eta)U}{C_C}. \tag{11}$$

It is ordinary for Eq. (11) to hold in the outsourcing setting, as $U >> C_C$.
- **Condition 2:** the server does not gain more payoff by switching to the action *Cheat*, i.e.,

$$p^{Server}_{(Verify,Not\ Cheat)} \geq p^{Server}_{(Verify,Cheat)}. \tag{12}$$

The server's payoff for the action profile *(Verify, Cheat)* depends on the catching probability ϵ. Formally,

$$p^{Server}_{(Verify,Cheat)} = \epsilon(-P_T - \beta C_S) + (1-\epsilon)(P - \beta C_S). \tag{13}$$

Let n_{FP} be the number of FP-strings, then ϵ measures the probability that among the αn_{FP} FP-strings for which the client calculates the edit distance, there exists at least one string that the server cheats on. In specific, we have

$$\epsilon = \begin{cases} 1 - \frac{\binom{n_{FP}}{\beta n_{FP}}\binom{n_{FP}-\beta n_{FP}}{\alpha n_{FP}}}{\binom{n_{FP}}{\alpha n_{FP}}\binom{n_{FP}}{\beta n_{FP}}} & \text{if } \alpha + \beta < 1 \\ 1 & \text{otherwise.} \end{cases} \tag{14}$$

To let Eq. (12) hold, it must be true that

$$\epsilon \geq \frac{(1-\beta)C_S}{P_T + P}. \tag{15}$$

After doing the inference, we have

$$\alpha \geq \frac{a_{min}}{n_{FP}}, \tag{16}$$

where $a_{min} = min\{a | \prod_{i=0}^{a-1} \frac{n_{FP}-\beta n_{FP}-i}{n_{FP}-i} \leq 1 - \frac{(1-\beta)C_S}{P_T+P}\}$.

In summary, by designing a strategic game as per Definition 7 between the client and the server, and by letting the client randomly picking at least α fraction of FP-strings as the V-strings for edit distance calculation, the server always chooses not to cheat on any FP-string to receive the maximum payoff.

The verification preparation phase and VO construction phase of P-VS^2 are exactly the same as G-VS^2. The only difference lies in **Step 3** of the VO verification phase. Instead of calculating the edit distance $DST(s, s_q)$ for every

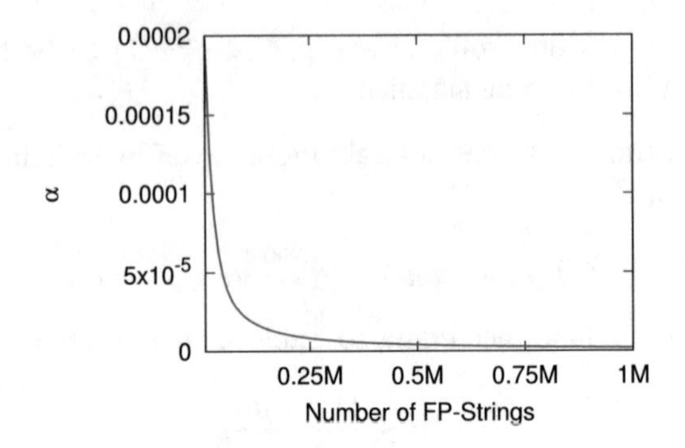

Fig. 7. α w.r.t. various n_{FP} ($\beta = 0.8$, $\eta = 0.9$, $C_S = 1$, $P = 10$, $P_T = 100$)

FP-string s in VO, the client only randomly picks α fraction of them as V-strings for verification. Therefore, P-VS^2 reduces the verification cost at the client side, especially when α is small. In Fig. 7, we show α with regard to various number of FP-strings. We observe that in all circumstances, α is no larger than 0.0002. More importantly, with the increase of n_{FP} (the number of FP-strings), α becomes even smaller. This result demonstrates that P-VS^2 works extremely well to overcome the drawback of G-VS^2 when there are plentiful FP-strings.

5 Complexity Analysis and Comparison

In this section, we compare the time and space complexity of the four approaches proposed in Sect. 4, namely VS^2 (Sect. 4.2), E-VS^2 (Sect. 4.3), G-VS^2 (Sect. 4.4) and P-VS^2 (Sect. 4.5). In specific, we analyze the time complexity of: (1) the verification preparation phase at the data owner side, (2) the VO construction phase at the server side, and (3) the VO verification phase at the client side. In addition, we present the space complexity of the VO, which impacts the bandwidth overhead between the server and the client.

In order to make the presentation clear, we first introduce the difference of the four approaches in Fig. 8. In every approach, to verify the soundness of M^S, the client needs to calculate the edit distance for every M-string. A naive approach to verify the completeness is that for each false hit, the client also calculates the edit distance. However, the induced verification cost is unacceptable, as the verification complexity is the equivalent to executing record matching locally. In VS^2, we optimize the completeness verification procedure by constructing a small number of MF-nodes for the NC-strings from the MB^{ed}-tree. In the verification phase, the client only needs to calculate the minimum edit distance between the target string s_q and every MF-node. We further reduce the verification complexity in E-VS^2. From those false hits that are not NC-strings, namely C-strings, we find a subset that are dissimilar to s_q in the Euclidean space, and call them E-strings. The server builds a small number of DBHs from the E-strings and includes them in the VO. In this way, by calculating quite

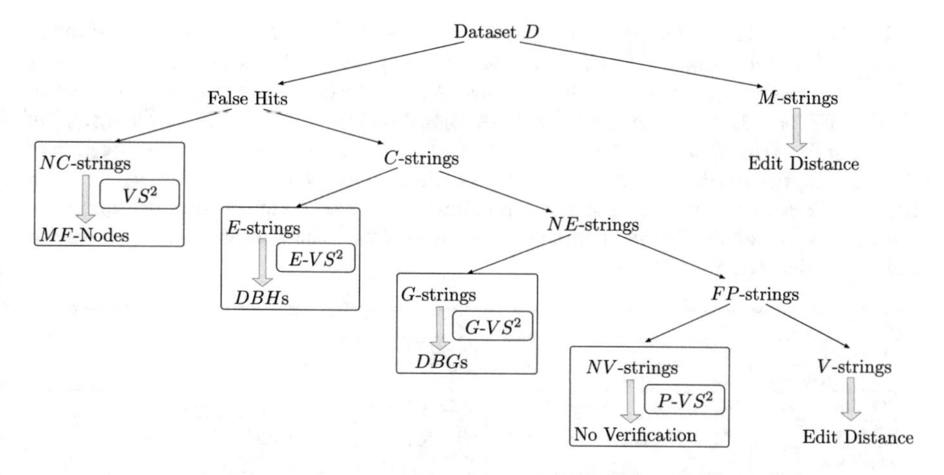

Fig. 8. Design comparison of VS^2, E-VS^2, G-VS^2 and P-VS^2

a few cheap Euclidean distance, the client learns that none of the E-strings are similar to s_q. Next, we propose G-VS^2 to reduce the verification cost for NE-strings. From the GC-vectors, the client is assured that every G-string is dissimilar to s_q. To boost the verification procedure, the server organizes plenty of G-strings into a few DBGs, so that the client reduces the amount of GC-vector distance calculation. Lastly, in P-VS^2, for those false hits that are left over, i.e., FP-strings, we only require the client to calculate the edit distance for a small portion of them. Table 3 summarizes the complexity for each approach.

In the verification preparation phase, in the VS^2 approach, the data owner constructs the MB^{ed}-tree with $O(n)$ complexity. While in E-VS^2, besides MB^{ed}-tree, the data owner needs to embed the strings into Euclidean points with $O(cdn^2)$ complexity. In both G-VS^2 and P-VS^2, the data owner needs to generate the GC-vector and RSA signature for each string. This process introduces an additional $O(n)$ complexity. However, the overall complexity is dominated by embedding.

Regarding the VO construction overhead at the server side, in VS^2, the server constructs VO by traversing the MB^{ed}-tree only. Therefore, the complexity is $O(n)$. While in E-VS^2, the DBH construction introduces $O(n_E^3)$ complexity. This makes the VO construction complexity of E-VS^2 higher than that of VS^2. In both G-VS^2 and P-VS^2, the server needs to execute Algorithm 1 to construct the DBGs, which adds on $O(n_G^2)$ complexity. However, the computing capability at the server side is solid. Thus the VO construction complexity of E-VS^2, G-VS^2 and P-VS^2 is still acceptable.

Before discussing the complexity of verification time, we must note that $n_C = n_E + n_{NE}$ and $n_{NE} = n_G + n_{FP}$, where n_C, n_E, n_{NE}, N_G and n_{FP} are the number of C-strings, E-strings, NE-strings, G-strings and FP-strings respectively. Usually, $n_{DBH} << n_E$ as a single DBH can cover the Euclidean points of a large number of E-strings. Also because a DBG embraces quite a few

Table 3. Complexity comparison between VS^2, $G\text{-}VS^2$ and $P\text{-}VS^2$ (c: a constant in [0, 1]; d: # of dimensions of Euclidean space; σ_S: Avg. length of the string; σ_M: size of a MB^{ed}-tree node; σ_D: size of a DBH; σ_H: size of a hash value; σ_{GC}: size of a GC-vector; σ_{DBG}: size of a DBG; σ_{sig}: size of a RSA signature; n: # of strings in D; n_R: # of strings in M^S; n_C: # of C-strings; n_{MF}: # of MF nodes; n_E: # of E-strings; n_{NE}: # of NE-strings; n_{DBH}: # of DBHs; n_G: # of G-strings; n_{FP}: # of FP-strings; n_{DBG}: # of DBGs; n_V: # of V-strings; C_{Ed}: the complexity of an edit distance computation; C_{El}: the complexity of Euclidean distance calculation; C_{GC}: the complexity of GC-vector distance calculation.)

Approach	VS^2	$E\text{-}VS^2$
Preparation	$O(n)$	$O(cdn^2)$
VO Construction	$O(n)$	$O(n + n_E^3)$
VO Verification	$O((n_R + n_{MF} + n_C)C_{Ed})$	$O((n_R + n_{MF} + n_{NE})C_{Ed} + n_{DBH}C_{El})$
VO Size	$O((n_R + n_C)\sigma_S + n_{MF}\sigma_M)$	$O((n_R + n_C)\sigma_S + n_{MF}\sigma_M + n_{DBH}\sigma_D)$
Approach	$G\text{-}VS^2$	$P\text{-}VS^2$
Preparation	$O(cdn^2)$	$O(cdn^2)$
VO Construction	$O(n + n_E^3 + n_G^2)$	$O(n + n_E^3 + n_G^2)$
VO Verification	$O((n_R + n_{MF} + n_{FP})C_{Ed} + n_{DBH}C_{El} + n_{DBG}C_{GC})$	$O((n_R + n_{MF} + n_V)C_{Ed} + n_{DBH}C_{El} + n_{DBG}C_{GC})$
VO Size	$O((n_R + n_C)\sigma_S + n_{MF}\sigma_M + n_{DBH}\sigma_D + n_G\sigma_{GC} + n_{DBG}\sigma_{DBG} + \sigma_{sig})$	$O((n_R + n_C)\sigma_S + n_{MF}\sigma_M + n_{DBH}\sigma_D + n_G\sigma_{GC} + n_{DBG}\sigma_{DBG} + \sigma_{sig})$

G-strings, we have $n_{DBG} << n_G$. Our empirical study shows that on a dataset of 1 million records, $\frac{n_E}{n_{DBH}} \approx 26$, and $\frac{n_G}{n_{DBG}} \approx 75$. Also note that C_{Ed} (i.e., complexity of an edit distance computation) is much more expensive than C_{El} and C_{GC} (i.e., the complexity of Euclidean distance and GC-vector distance calculation). Our experiments show that the time to compute one single edit distance can be 20 times of computing one Euclidean distance and 150 times of GC-vector distance. Therefore, compared with VS^2, $E\text{-}VS^2$ significantly reduces the verification overhead at the client side, as it substitutes n_E edit distance calculation for E-strings with a small amount of Euclidean distance calculation for the DBHs. Analogically, on top of $E\text{-}VS^2$, $G\text{-}VS^2$ alleviates the verification cost as it replaces the edit distance calculation for the G-strings with the GC-vector distance computation for a few DBGs. $P\text{-}VS^2$ requires least efforts from the client, because it only requires the client to calculate edit distance for V-strings, which is only a small fraction (α) of the FP-strings.

Regarding the VO size, the VO size of the VS^2 approach is calculated as the sum of two parts: (1) the total size of the similar strings and C-strings (in string format), and (2) the size of MF nodes. Note that $\sigma_M = 2\sigma_S + \sigma_H$, where σ_H is the size of a hash value. In our experiments, it turned out that $\sigma_M/\sigma_S \approx 10$. The VO size of the $E\text{-}VS^2$ approach is calculated as the sum of three parts: (1) the total size of the similar strings and C-strings (in string format), (2) the size of MF nodes, and (3) the size of DBHs. In addition to the VO of $E\text{-}VS^2$, the VO of the $G\text{-}VS^2$ and $P\text{-}VS^2$ approaches have two additional parts: the GC-vectors of G-strings and a Condensed-RSA signature for all the G-strings.

Our experimental results show that σ_D and σ_{GC} is small. Hence, the VO size of these four approaches are comparable.

In summary, compared with the basic approach VS^2, $E\text{-}VS^2$, $G\text{-}VS^2$ and $P\text{-}VS^2$ dramatically reduce the verification complexity at the client side at the expense of more preparation overhead at the data owner side and more VO construction overhead at the server side. However, as the preparation procedure is executed only once, the cost of constructing the Euclidean points and *GC*-vectors can be amortized by a large number of queries from the client. And as the server is equipped with strong computing resources, it is of the highest priority to reduce the verification complexity at the client side. Therefore, $E\text{-}VS^2$ and $G\text{-}VS^2$ are more appropriate for the authentication of similarity search on outsourced databases. $P\text{-}VS^2$ decreases the verification cost even further, but it demands a strategic game between the client and server. This approach is more suitable for clients with extremely limited computational resources.

6 Experiments

6.1 Experiment Setup

Datasets and Queries. We use two real-world datasets: (1) the *Actors* dataset from IMDB[3] that contains 260 K last names. The maximum length of a name is 63, while the average is 14.627; and (2) the *Authors* dataset from DBLP[4] including 1,000,000 researcher names. The maximum length of a name is 99 while the average length is 14.732. As both datasets are very large, it is too time-consuming to execute approximate matching for all records in these two datasets. Thus, we picked 10 target strings for approximate matching. For the following results, we report the average performance of matching of the 10 target strings.

Experimental Environment. We implement the four approaches, namely VS^2, $E\text{-}VS^2$, $G\text{-}VS^2$ and $P\text{-}VS^2$, in C++. The hash function we use is the *SHA256* function from the *OpenSSL library*[5]. We use the *gmpxx* interface of the *GMP library*[6] to implement the *Condensed-RSA* module. We execute the experiments on a machine with 2.4 GHz CPU and 48 GB RAM, running Linux 3.2.

Baseline. As the baseline method, we implement the brute-force (BF) approximate string matching algorithm that calculates the edit distance between every string in the dataset and the target string s_q. Additionally, we compare our approaches against the state-of-the-art similarity search approach named B^{ed}-tree Zhang et al. (2010). We do not take the Merkle-tree based verification R.C.Merkle (1980) as a baseline, since it only verifies authenticity.

[3] http://www.imdb.com/interfaces.
[4] http://dblp.uni-trier.de/xml/.
[5] https://www.openssl.org/.
[6] https://gmplib.org/.

Parameter Setup. The parameters include: (1) string edit distance threshold θ, and (2) the fanout f of MB^{ed}-tree nodes (i.e., the number of entries that each node contains). The details of the parameter settings can be found in Table 4. We apply *SparseMap* Hjaltason and Samet (2003) to embed every string into a 5-dimensional point in the Euclidean space. Also, in $G\text{-}VS^2$, we convert each string to a GC-vector of length 4.

Table 4. Parameter settings

Parameter	Setting
Similarity threshold θ	2,3,4,5,6,7
MB^{ed}-tree fanout f	10, 100, 1000, 10000
Embedding dimension d	5
GC-vector length L	4

6.2 VO Construction Time

First, we measure the impact of distance threshold θ on the VO construction time. In Fig. 9, we show the VO construction time with regard to different θ values on both datasets. It is worth noting that $G\text{-}VS^2$ and $P\text{-}VS^2$ share the same VO construction process. Therefore, we plot the time performance of $G\text{-}VS^2$ and $P\text{-}VS^2$ in the same line. First, we observe that $\text{E-}VS^2$ consumes more time on constructing VOs than VS^2, especially when θ is a small value. This is because besides traversing the MB^{ed}-tree like VS^2, $\text{E-}VS^2$ constructs DBHs for the E-strings. The number of E-strings n_E is large when θ is small. For example, on the *Authors* dataset, when $\theta = 1$, $n_E = 510,185$; while when $\theta = 6$, $n_E = 111,129$. As a result, the DBH construction time is higher when the threshold is small. Second, the VO construction time of $G\text{-}VS^2$ and $P\text{-}VS^2$ is also large when θ is small. This is because the large number of G-strings bring about more DBG construction time. Another interesting observation is

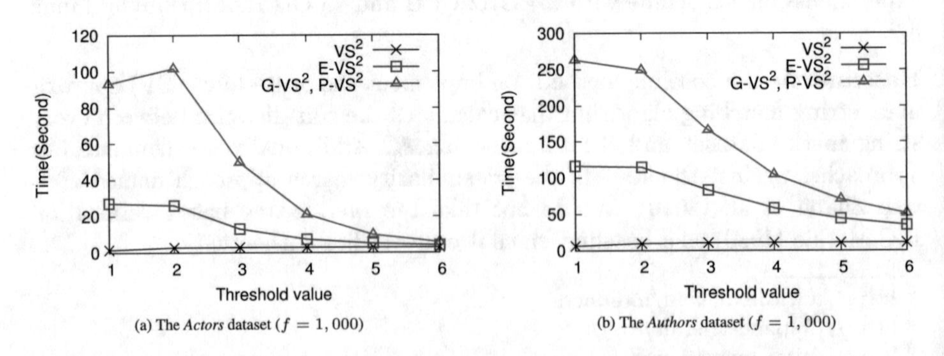

(a) The *Actors* dataset ($f = 1,000$)
(b) The *Authors* dataset ($f = 1,000$)

Fig. 9. VO construction time w.r.t. distance threshold θ

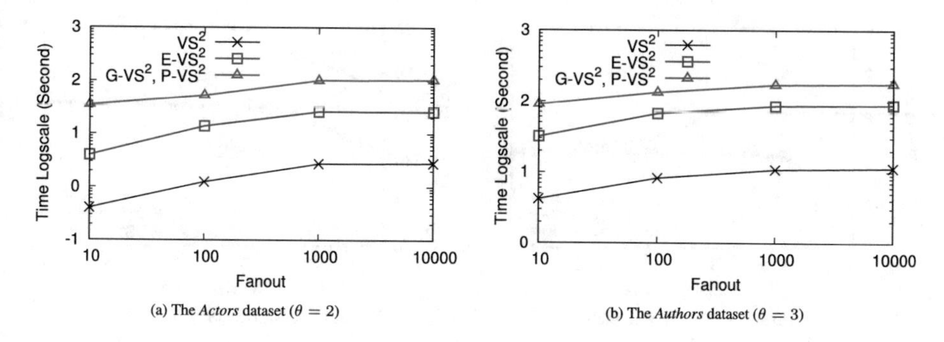

(a) The *Actors* dataset ($\theta = 2$)

(b) The *Authors* dataset ($\theta = 3$)

Fig. 10. VO construction time w.r.t. MB^{ed}-tree fanout f

that on the both datasets, the VO construction time is long when $\theta = 2$. We further investigate the reason and find that the number of G-strings n_G reaches its maximum value when $\theta = 2$. Third, the VO construction time of VS^2 keeps stable with the variation of θ. As in VS^2, the server only traverses the MB^{ed}-tree for VO construction, the change in θ has limited impact on the number of nodes to be visited. Fourth, in general, there is a dramatic decrease in VO construction time of E-VS^2, G-VS^2 and P-VS^2 when θ increases. The reason lies in the sharp reduction in n_E and n_G. Since the complexity of DBH construction and DBG construction are $O(n_E^3)$ and $O(n_G^2)$ respectively, the total VO construction time decreases intensively when n_E and n_G decrease. Last, the VO construction time on the *Authors* dataset is larger than that on the *Actors* dataset. This is because with the same fanout f, the MB^{ed}-tree's size of the *Authors* dataset is larger, due to the greater data size.

We also measured the impact of the MB^{ed}-tree structure on the VO construction time. From Fig. 10, we observe that compared with VS^2, E-VS^2 and G-VS^2 (P-VS^2) add the VO construction cost by a magnitude of 1 and 2 respectively, under various fanout of the MB^{ed}-tree. The underlying reason resides in the additional DBH construction cost of E-VS^2, and the DBG construction cost of G-VS^2 (P-VS^2). Besides, in all the experiments, the VO construction time increases with the fanout value f. Even though the total number of nodes in the MB^{ed}-tree decreases with larger fanout values, the decrease in the number of MF-nodes n_{MF} is more intense. For example, on the *Authors* dataset, when $f = 100$, $n_{MF} = 5,420$; while when $f = 1,000$, $n_{MF} = 57$. The reason behind the significant reduction in n_{MF} is that when f is large, $DST_{min}(s_q, N)$ becomes a loose lower-bound for any MB^{ed}-tree node N. This leads to a significant increase in the number of E-strings and G-strings. The observation from Fig. 10 suggests that small fanout values fit our verification approaches better.

6.3 VO Size

We measure the impact of the distance threshold θ on the VO size. The results are shown in Fig. 11. We notice that the VO size of VS^2 and E-VS^2 is very

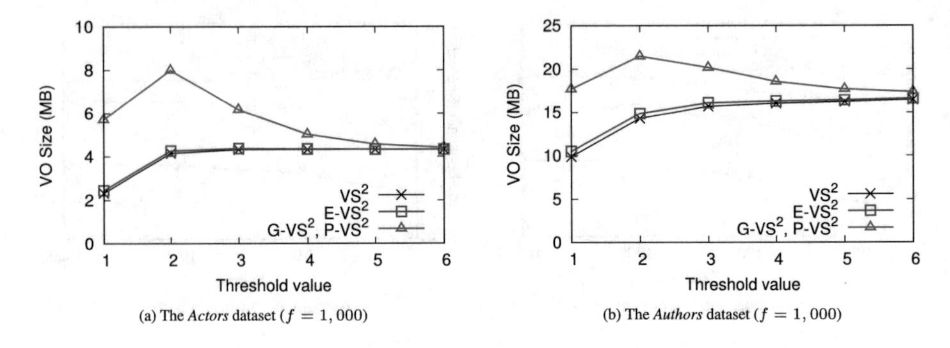

(a) The *Actors* dataset ($f = 1,000$)

(b) The *Authors* dataset ($f = 1,000$)

Fig. 11. VO size w.r.t. distance threshold θ

close. The reason is two manifold. On the one hand, the size of each DBH is very small. On the other hand, the number of DBHs n_{DBH} is also very small. In the experiments, $n_E/n_{DBH} \approx 26$. In other words, each DBH covers the Euclidean points of 26 E-strings. Due to these two reasons, even though E-VS^2 introduces DBHs additionally to the VO, the increment in VO size is small, and the overall VO size is roughly the same as that of VS^2. Another observation is that in E-VS^2, the VO size increases with the growth of θ value, especially when θ increases from 1 to 2. Apparently, larger θ values lead to more similar strings (i.e., larger n_R), fewer MF-nodes (i.e., smaller n_{MF}), and fewer C-strings (i.e., smaller n_C). In VS^2, the VO size is decided by $(n_R + n_C)\sigma_S + n_{MF}\sigma_M$, where $\frac{\sigma_M}{\sigma_S} \approx 20$. When we increase θ, even though n_{MF} reduces, the intensive growth of n_R and n_C leads to the increase of VO size. For example, on the *Actors* dataset, when $\theta = 1$, $n_R + n_C = 137,655$, $n_{MF} = 247$. While when $\theta = 3$, $n_R + n_C = 249,373$, $n_{MF} = 22$. However, compared with E-VS^2, G-VS^2 and P-VS^2 introduce more space overhead in the VO. This is because the server needs to insert the GC-vector for each G-string into the VO. We also notice that on both datasets, the VO size is the largest when $\theta = 2$. The reason is that the number of G-strings is at the top when $\theta = 2$. For instance, on the *Actors* dataset, when $\theta = 1, 2, 3$, the number of G-strings are 116,209, 133,059 and 64,284 respectively.

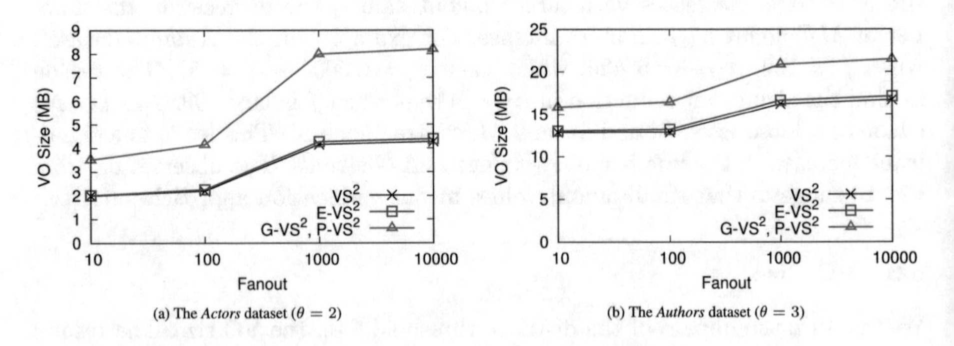

(a) The *Actors* dataset ($\theta = 2$)

(b) The *Authors* dataset ($\theta = 3$)

Fig. 12. VO size w.r.t. MB^{ed}-tree fanout f

Furthermore, we measure the impact of the MB^{ed}-tree fanout f on the VO size. Following the same reason as before, we notice in Fig. 12 that VS^2 and E-VS^2 produce VO of similar size, while G-VS^2 and P-VS^2 construct VO of larger size. For all approaches, the VO size increases with the growth of f. Recall that in VS^2, the VO size is $(n_R + n_C)\sigma_S + n_{MF}\sigma_M$. When f increases, n_R is unchanged. Meanwhile, n_C and n_G dramatically increases with f (e.g., on the *Actors* dataset, when f increases from 100 to 1,000, n_C and n_G increase by 122,515 and 33,469 respectively). Moreover, when f increases, n_{MF} decreases by a small degree (when f changes from 100 to 1,000, n_{MF} decreases by 2,937). Considering that $\frac{\sigma_M}{\sigma_S} \approx 20$, the intensive increase of n_C and n_G lead to a larger VO size.

6.4 VO Verification Time

In this section, we measure the VO verification time at the client side. We compare the VO verification time with the performance of our baseline algorithms, aiming to show that the client's verification effort is less than by doing string matching locally.

In Fig. 13, we report the VO verification time of our four approaches, as well as the baselines, with regard to various distance threshold values. First, we observe that the VO verification time is always smaller than BF. In particular, it can save up to 53% computational effort at the client side if she verifies the approximate matching result using VS^2 rather than using BF. P-VS^2 can save even 98% of the client's verification effort. E-VS^2, G-VS^2 and P-VS^2 are more efficient than executing record matching on a B^{ed}-tree. The advantage is more evident when θ is small. This proves that our verification method is suitable for the outsourcing setting. Second, the verification time of VS^2, E-VS^2 and G-VS^2 increase with the growth of θ. This is because in general, when θ increase, the number of MF-nodes, E-strings and G-strings constantly decrease, which decrements the power of saving verification time of these approaches. As a contrast, the verification time of P-VS^2 keeps stable with the change of θ. The underlying reason is that P-VS^2 significantly reduces the number of edit distance

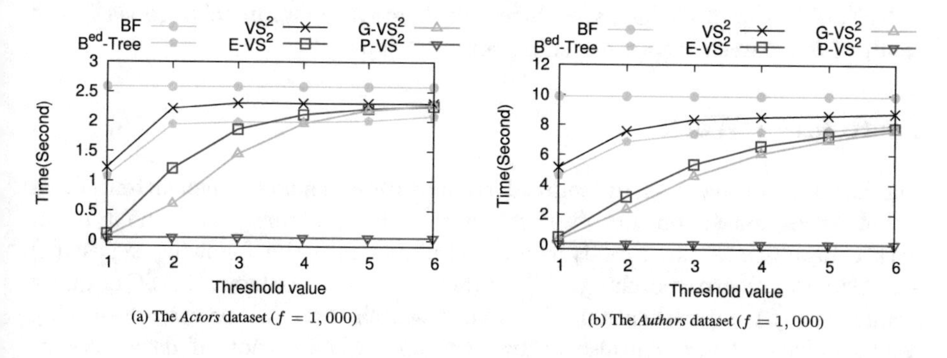

(a) The *Actors* dataset ($f = 1,000$)

(b) The *Authors* dataset ($f = 1,000$)

Fig. 13. VO verification time w.r.t. distance threshold θ

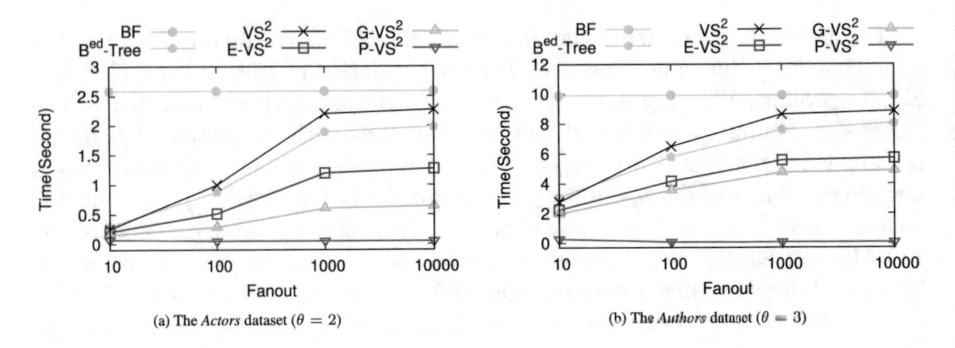

(a) The *Actors* dataset ($\theta = 2$)

(b) The *Authors* dataset ($\theta = 3$)

Fig. 14. VO verification time w.r.t. MB^{ed}-tree fanout f

computation in the verification procedure. For instance, on the *Authors* dataset, when $\theta = 3$, the client only needs to calculate 216 edit distances in total. This makes the modular exponentiation in Eq. (9) dominate the verification time expense in $P\text{-}VS^2$. As a consequence, we observe constant verification time performance. Third, in accordance with our theoretical analysis in Sect. 5, our four approaches accumulatively save the verification time at the client side. With $P\text{-}VS^2$, for a dataset that includes 1 million records, the client is able to finish the verification of the record matching result within 0.05 s (Fig. 14).

We also measure the VO verification time with various MB^{ed}-tree fanout f. On both datasets, in most cases, the VO verification is always more efficient than the baseline methods. In particular, when the fanout reaches 10, compared with BF, E-VS^2 and G-VS^2 can save up to 92% and 97% computational effort at the client side respectively. Besides, the verification time increases when f grows. This is straightforward as larger f results in the reduction of the number of NC-strings for VS^2, E-strings for E-VS^2 and G-strings for $G\text{-}VS^2$. This requires more edit distance calculations. While as the verification time of $P\text{-}VS^2$ keeps constant. The reason is similar to the previous discussion. Even though there is a huge number of FP-strings, $P\text{-}VS^2$ only asks to calculate the edit distance for an extremely fraction of them.

In sum, the experiment results suggest that our four approaches, namely VS^2, $E\text{-}VS^2$, $G\text{-}VS^2$ and $P\text{-}VS^2$, are efficient authentication approaches to verify the result of outsourced record matching computations.

7 Related Work

The security problem of outsourced computations caught much attention in recent years. Based on the perspective to enforce security and the type of data computations, we classify these techniques into the following types: (1) approximate record matching, (2) authentication of outsourced SQL query evaluation, (3) authentication of keyword search, and (4) privacy-preserving record linkage. There are also other work on authentication of data integration Chen et al. (2015) and in-network aggregation in distributed systems (e.g. Papadopoulos et al. (2012); Zhang et al. (2013a)). None of these work considered the authentication of outsourced approximate string matching.

Approximate Record Matching. Record matching (or record linkage, similarity join) aims at linking records from different sources based on their similarity. One popular similarity measurement metric is edit distance. Due to the quadric complexity of edit distance calculation, numerous pruning approaches have been proposed to reduce the amount of necessary calculations. MinHash Cohen et al. (2001) was proposed to construct digests from records in a similarity-preserving way. For any pair of records, if they are similar (i.e., have small edit distance), highly likely their digests are similar. The authors in Cohen et al. (2001) defined a function to generate the candidate matching pairs from the digests efficiently. Bronstein et al. (2010) proposed WaldHash to solve the record matching problem in a sequential decision strategy by designing a sequential similarity-preserving hashing. Park et al. (2015) designed the neighbor-sensitive hashing to facilitate the nearest neighbor search from the digests. StringMap was proposed in Jin et al. (2003) to embed strings into Euclidean space in a similarity-preserving way. By efficiently discovering a small amount of pivot strings from a large dataset, any string can be embedded as an Euclidean point. A small set of candidate matching pairs can be discovered from the Euclidean space for further investigation. Gravano et al. (2001) inferred the relationship between the edit distance and the number of mismatching q-grams between two strings. With that reasoning, for any pair of strings, if the number of mismatching q-grams is larger than a certain value, their edit distance must be larger than the threshold value. Therefore, this blocking technique helps to filter out many candidate strings. Xiao et al. (2008) derived a tighter edit distance lower bound by analyzing the number of mismatches of the locational q-grams of two strings. Following that, a new algorithm named *Ed-Join* is designed to boost the join efficiency. Salmela and Tarhio Salmela and Tarhio (2008) extended the q-gram based blocking techniques to parameterized string matching. In this paper, we propose an efficient approach for the client to get approximate record matching results and to obtain correctness confidence in the results.

Authentication of Outsourced SQL Query Evaluation. The issue of providing authenticity for outsourced database was initially raised in the database-as-a-service (DaS) paradigm Hacigümüş et al. (2002). The aim is to assure the correctness of SQL query evaluation over the outsourced databases. The proposed solutions include Merkle hash trees Li et al. (2006); Mykletun et al. (2006), signatures on a chain of paired tuples Pang et al. (2009), and authenticated B-tree and R-tree structures for aggregated queries Li et al. (2010). The key idea of these techniques is that the data owner outsources not only data but also the endorsements of the data being outsourced. These endorsements are signed by the data owner against tampering with by the service provider. For the cleaning results, the service provider returns both the results and a proof, called the *verification object* (VO), which is an auxiliary data structure to store the processing traces such as index traversals. The client uses the VO, together with the answers, to reconstruct the endorsements and thus verify the authenticity of the results. An efficient authentication technique should minimize the size of VO, while requiring lightweight authentication at the client side. Bajaj et al. Bajaj and Sion (2013) propose *CorrectDB* by incorporating trusted hardware and cryptographic techniques. It supports the authentication of arbitrary SQL queries over the outsourced database. In this paper, we follow the same VO-based strategy to design our authentication method. But we consider record matching as the computation to be authenticated, rather than SQL queries.

Authentication of Keyword Search. Pang et al. Pang and Mouratidis (2008) targeted at search engines that perform similarity-based document retrieval, and designed a novel authentication mechanism for the search results. The key idea of the authentication is to build the Merkle hash tree (MHT) on the inverted index, and use the MHT for VO construction. Though effective, it has several limitations, e.g., the MHT cannot deal data updates efficiently Goodrich et al. (2012). Goodrich et al. (2012) designed a new model that considered conjunctive keyword searches as equivalent with a set intersection on the underlying inverted index data structure. They use the authenticated data structure in Papamanthou et al. (2011) to verify the correctness of set operations. Sun et al. (2014) considered cosine similarity measurement, and designed encrypted data search functionality that support multi-keyword queries, result ranking and result verification. Its search index is built based on the vector space model. By combining the suffix tree with set difference verification protocol, Papadopoulos et al. (2015) verifies the outsourced pattern matching result with minimum VO size. In this paper, we augment the existing work by facilitating the authentication of approximate keyword search.

Privacy-preserving Record Linkage. The privacy of sensitive information in the outsourced record matching has been the focus of research for two decades. Schnell et al. (2009) and Durham et al. (2014) propose to store the q-grams in the Bloom Filters (BFs); the BFs of two similar strings (i.e., many q-grams in common) have a large number of identical bit positions set to 1. However, Schnell et al. (2009) observed that the BFs are still open to the frequency attack. Storer et al. (2008) consider the duplicates as exact identical contents. They use convergent encryption that uses a function of the hash of the plaintext of a chunk as the encryption key, so that any identical plaintext values will be encrypted to identical ciphertext values. However, the convergent encryption disables the approximate matching on outsourced records. Bonomi et al. (2012) propose a new transformation method that integrates embedding methods with differential privacy. Its embedding strategy projects the original data on a base formed by a set of frequent variable length grams. The privacy of the gram construction procedure is guaranteed to satisfy differential privacy. Atallah et al. (2003) design a SMC protocol based on homomorphic encryption for record linkage. Zhang et al. (2013b) consider the potential collusions between parties that execute the record matching. By exploiting cryptographic operations such as pseudorandom permutations, a collusion-resistant protocol is proposed to protect the data privacy. In this paper, we address the security issue of outsourced record matching from a different perspective by providing result correctness verification approaches.

8 Conclusion

In this paper, we designed an efficient authentication framework, namely *ALARM*, for outsourced approximate record matching. *ALARM* consists of four authentication approaches, i.e., VS^2, E-VS^2, G-VS^2 and P-VS^2. We formally prove the robustness and security of these approaches. We also theoretically analyze the time and space complexity of these approaches in each phase of authentication. Experimental results show that our approaches can authenticate the record matching result efficiently.

There are several interesting research directions to explore in the future. First, in this paper, we only focus on record matching for outsourced databases based on string edit distance. In the future, we plan to extend the work to support authentication of the matching results based on string semantic similarity. Second, we acknowledge the application of machine learning techniques, such as convolutional neural network Gottapu et al. (2016); Wilson (2011) and support vector machine (SVM) Feigenbaum (2016), that are leveraged for record linkage. The verification approach designed in this paper relies on a specific similarity measurement and thus cannot be adapted to these machine learning techniques. In the future, we plan to investigate the authentication method for the machine learning based record linkage results. Third, we plan to provide statistical correctness evaluation of the matching results returned by the server, such as precision and recall. Our current authentication methods only can return a binary decision (YES/NO) for the incorrectness/incompleteness. Providing the quantitative information of correctness/completeness such as precision and recall can help the client pick the server of best service quality when there are multiple servers that provide similar services. However, it is challenging to provide precision and recall of the returned results, as it requires the ground truth of matching, which is difficult for the client to obtain, due to its limited computational resources. In the future, we will explore the lightweight verification methods that can estimate the precision and recall of the returned matching results.

Acknowledgements. This material is based upon work supported by the National Science Foundation (NSF) under Grant No. 1350324 and 1464800.

References

Arasu, A., Ganti, V., Kaushik, R.: Efficient exact set-similarity joins. In: Proceedings of the VLDB Endowment (2006)

Atallah, M.J., Kerschbaum, F., Du, W.: Secure and private sequence comparisons. In: Proceedings of the Workshop on Privacy in the Electronic Society (2003)

Bajaj, S., Sion, R.: CorrectDB: SQL engine with practical query authentication. Proc. VLDB Endow. **6**(7), 529–540 (2013)

Beckmann, N., Kriegel, H.P., Schneider, R., Seeger, B.: The R*-tree: an efficient and robust access method for points and rectangles. In: ACM Sigmod Record, vol. 19, pp. 322–331. ACM (1990)
Bonomi, L., Xiong, L., Chen, R., Fung, B.: Frequent grams based embedding for privacy preserving record linkage. In: Proceedings of the International Conference on Information and Knowledge Management (2012)
Bronstein, A.M., Bronstein, M.M., Guibas, L.J., Ovsjanikov, M.: Waldhash: sequential similarity-preserving hashing. Technical report CIS-2010-03, Technion, Israel (2010)
Chaudhuri, S., Ganjam, K., Ganti, V., Motwani, R.: Robust and efficient fuzzy match for online data cleaning. In: Proceedings of the International Conference on Management of Data (2003)
Chen, Q., Hu, H., Xu, J.: Authenticated online data integration services. In: Proceedings of the International Conference on Management of Data (2015)
Cohen, E., Datar, M., Fujiwara, S., Gionis, A., Indyk, P., Motwani, R., Ullman, J.D., Yang, C.: Finding interesting associations without support pruning. IEEE Trans. Knowl. Data Eng. **13**(1), 64–78 (2001)
Comer, D.: Ubiquitous b-tree. Comput. Surv. (1979)
Cormen, T.H.: Introduction to Algorithms. MIT press, Cambridge (2009)
De Lathauwer, L., De Moor, B., Vandewalle, J.: A multilinear singular value decomposition. SIAM J. Matrix Anal. Appl. **21**(4), 1253–1278 (2000)
Dong, B., Wang, H.: Efficient authentication of outsourced string similarity search. CoRR abs/1603.02727 (2016). http://arxiv.org/abs/1603.02727
Dong, B., Liu, R., Wang, W.H.: PraDa: Privacy-preserving data-deduplication-as-a-service. In: Proceedings of the International Conference on Conference on Information and Knowledge Management (2014)
Draper, N.R., Smith, H.: Applied Regression Analysis. Wiley, New York (2014)
Durham, E.A., Kantarcioglu, M., Xue, Y., Toth, C., Kuzu, M., Malin, B.: Composite bloom filters for secure record linkage. Trans. Knowl. Data Eng. (2014)
Eidenbenz, S., Stamm, C.: Maximum clique and minimum clique partition in visibility graphs. In: Theoretical Computer Science, pp. 200–212 (2000)
Faloutsos, C., Lin, K.I.: FastMap: A Fast Algorithm for Indexing, Data-mining and Visualization of Traditional and Multimedia Datasets, vol. 24 (1995)
Feigenbaum, J.J.: A machine learning approach to census record linking (2016). Accessed 28 Mar 2016
Goodrich, M.T., Papamanthou, C., Nguyen, D., Tamassia, R., Lopes, C.V., Ohrimenko, O., Triandopoulos, N.: Efficient verification of web-content searching through authenticated web crawlers. Proc. VLDB Endow. (2012)
Gottapu, R.D., Dagli, C., Ali, B.: Entity resolution using convolutional neural network. Proc. Comput. Sci. **95**, 153–158 (2016)
Gravano, L., Ipeirotis, P.G., Jagadish, H.V., Koudas, N., Muthukrishnan, S., Srivastava, D., et al.: Approximate string joins in a database (almost) for free. In: VLDB, vol. 1, pp. 491–500 (2001)
Guha, S., Mishra, N.: Clustering data streams. In: Data Stream Management, pp 169–187. Springer (2016)
Hacigümüş, H., Iyer, B., Li, C., Mehrotra, S.: Executing SQL over encrypted data in the database-service-provider model. In: Proceedings of the International Conference on Management of Data (2002)
Hazay, C., Lewenstein, M., Sokol, D.: Approximate parameterized matching. ACM Trans. Algorithms (TALG) **3**(3), 29 (2007)
Hjaltason, G., Samet, H.: Contractive embedding methods for similarity searching in metric spaces. Technical report TR-4102, Computer Science Department (2000)

Hjaltason, G.R., Samet, H.: Properties of embedding methods for similarity searching in metric spaces. Trans. Pattern Anal. Mach. Intell. (2003)
Jin, L., Li, C., Mehrotra, S.: Efficient record linkage in large data sets. In: International Conference on Database Systems for Advanced Applications (2003)
Kamel, I., Faloutsos, C.: Hilbert r-tree: An improved r-tree using fractals. Technical report (1993)
Koudas, N., Sarawagi, S., Srivastava, D.: Record linkage: similarity measures and algorithms. In: Proceedings of the International Conference on Management of Data (2006)
Li, C., Lu, J., Lu, Y.: Efficient merging and filtering algorithms for approximate string searches. In: International Conference on Data Engineering (2008)
Li, F., Hadjieleftheriou, M., Kollios, G., Reyzin, L.: Dynamic authenticated index structures for outsourced databases. In: Proceedings of the International Conference on Management of Data (2006)
Li, F., Hadjieleftheriou, M., Kollios, G., Reyzin, L.: Authenticated index structures for aggregation queries. Trans. Inf. Syst. Secur. (2010)
Merkle, R.C.: Secure communications over insecure channels. Commun. ACM **21**(4), 294–299 (1978)
Miller, A., Hicks, M., Katz, J., Shi, E.: Authenticated data structures, generically. In: ACM SIGPLAN Notices, vol. 49, pp. 411–423 (2014)
Moon, B., Jagadish, H.V., Faloutsos, C., Saltz, J.H.: Analysis of the clustering properties of the hilbert space-filling curve. IEEE Trans. Knowl. Data Eng. **13**(1), 124–141 (2001)
Morris, P.: Introduction to game theory. Springer, New York (2012)
Morton, G.M.: A Computer Oriented Geodetic Data Base and a New Technique in File Sequencing. International Business Machines Company, New York (1966)
Mykletun, E., Narasimha, M., Tsudik, G.: Authentication and integrity in outsourced databases. ACM Trans. Storage (TOS) **2**(2), 107–138 (2006)
O'Connell, R.T., Koehler, A.B.: Forecasting, time series, and regression: An applied approach, vol. 4. South-Western Pub (2005)
Pang, H., Mouratidis, K.: Authenticating the query results of text search engines. Proc. VLDB Endow. (2008)
Pang, H., Zhang, J., Mouratidis, K.: Scalable verification for outsourced dynamic databases. Proc. VLDB Endow. (2009)
Papadopoulos, D., Papamanthou, C., Tamassia, R., Triandopoulos, N.: Practical authenticated pattern matching with optimal proof size. Proc. VLDB Endow. **8**(7), 750–761 (2015)
Papadopoulos, S., Wang, L., Yang, Y., Papadias, D., Karras, P.: Authenticated multistep nearest neighbor search. Trans. Knowl. Data Eng. (2011)
Papadopoulos, S., Kiayias, A., Papadias, D.: Exact in-network aggregation with integrity and confidentiality. Trans. Knowl. Data Eng. (2012)
Papamanthou, C., Tamassia, R.: Time and space efficient algorithms for two-party authenticated data structures. In: International Conference on Information and Communications Security, pp. 1–15. Springer (2007)
Papamanthou, C., Tamassia, R., Triandopoulos, N.: Optimal verification of operations on dynamic sets. In: Advances in Cryptology (2011)
Park, Y., Cafarella, M., Mozafari, B.: Neighbor-sensitive hashing. Proc. VLDB Endow. **9**(3), 144–155 (2015)
Parsons, S., Wooldridge, M.: Game theory and decision theory in multi-agent systems. Auton. Agents Multi-Agent Syst. **5**(3), 243–254 (2002)

Raimondi, F., Pecheur, C., Lomuscio, A.: Applications of model checking for multi-agent systems: verification of diagnosability and recoverability. In: Proceedings of Concurrency, Specification & Programming (CS&P), Warsaw University, pp. 433–444 (2005)

Ravikumar, P., Cohen, W.W., Fienberg, S.E.: A secure protocol for computing string distance metrics. In: the Workshop on Privacy and Security Aspects of Data Mining (2004)

Merkle, R.C.: Protocols for public key cryptosystems. In: Symposium on Security and Privacy (1980)

Rivest, R.L., Shamir, A., Adleman, L.: A method for obtaining digital signatures and public-key cryptosystems. Commun. ACM **21**(2), 120–126 (1978)

Salmela, L., Tarhio, J.: Fast parameterized matching with q-grams. J. Discrete Algorithms **6**(3), 408–419 (2008)

Schnell, R., Bachteler, T., Reiher, J.: Privacy-preserving record linkage using bloom filters. BMC Med. Inf. Decis. Making (2009)

Smola, A.J., Schölkopf, B.: A tutorial on support vector regression. Stat. Comput. **14**(3), 199–222 (2004)

Steinbach, M., Karypis, G., Kumar, V., et al.: A comparison of document clustering techniques. In: KDD Workshop on Text Mining, Boston, vol. 400, pp. 525–526 (2000)

Storer, M.W., Greenan, K., Long, D.D., Miller, E.L.: Secure data deduplication. In: Proceedings of the International Workshop on Storage Security and Survivability (2008)

Sun, W., Wang, B., Cao, N., Li, M., Lou, W., Hou, Y.T., Li, H.: Verifiable privacy-preserving multi-keyword text search in the cloud supporting similarity-based ranking. IEEE Trans. Parallel Distrib. Syst. **25**(11), 3025–3035 (2014)

Sutinen, E., Tarhio, J.: On using q-gram locations in approximate string matching. In: AlgorithmsESA 1995, pp. 327–340 (1995)

Turner V, Gantz JF, Reinsel D, Minton S (2014) The digital universe of opportunities: Rich data and the increasing value of the internet of things. IDC Analyze the Future

Ukkonen, E.: Approximate string-matching with q-grams and maximal matches. Theor. Comput. Sci. **92**(1), 191–211 (1992)

Wilson, D.R.: Beyond probabilistic record linkage: Using neural networks and complex features to improve genealogical record linkage. In: The International Joint Conference on Neural Networks, pp. 9–14. IEEE (2011)

Xiao, C., Wang, W., Lin, X.: Ed-join: an efficient algorithm for similarity joins with edit distance constraints. Proc. VLDB Endow. (2008)

Yang, Y., Papadias, D., Papadopoulos, S., Kalnis, P.: Authenticated join processing in outsourced databases. In: Proceedings of the International Conference on Management of Data (2009)

Zhang, R., Shi, J., Zhang, Y., Zhang, C.: Verifiable privacy-preserving aggregation in people-centric urban sensing systems. J. Sel. Areas Commun. (2013a)

Zhang, Y., Wong, W.K., Yiu, S.M., Mamoulis, N., Cheung, D.W.: Lightweight privacy-preserving peer-to-peer data integration. In: VLDB Endowment, pp. 157–168 (2013b)

Zhang, Z., Hadjieleftheriou, M., Ooi, B.C., Srivastava, D.: Bed-tree: an all-purpose index structure for string similarity search based on edit distance. In: Proceedings of the International Conference on Management of Data (2010)

Zimmer, M.: The twitter archive at the library of congress: challenges for information practice and information policy. First Monday **20**(7) (2015)

Active Dependency Mapping

A Data-Driven Approach to Mapping Dependencies in Distributed Systems

Alexia Schulz[1(✉)], Michael Kotson[1], Chad Meiners[1], Timothy Meunier[1], David O'Gwynn[2], Pierre Trepagnier[1], and David Weller-Fahy[1]

[1] Cyber Security and Information Sciences MIT Lincoln Laboratory, 244 Wood Street, Lexington, MA 02420, USA
alexia.schuz@ll.mit.edu

[2] Department of Computer Science, Belhaven University, 1500 Peachtree Street, Jackson, MS 39202, USA

Abstract. In this paper we introduce Active Dependency Mapping (ADM), a method for establishing dependency relations among a set of interdependent services. Consider the old expression among sysadmins: the way to discover who is using a server is to "turn it off and see who complains." ADM takes the same approach to dependency mapping. We directly establish dependency relationships between entities in an ecosystem of interacting services by instrumenting a user-facing application, and systematically blocking access to other entities, observing which blocks impact performance.

Our approach to ADM differs in an important way from "turning it off to see who complains." Although ADM depends on degrading the network environment, our goal is to minimize the impact of traffic blocks by reducing the duty cycle to determine the shortest block that enables dependency mapping. In an ideal case, dependencies may be established with no one complaining. Artificial degradation of the network environment could be transparent to users; run continuously it could identify dependencies that are rare or occur only at certain timescales. A useful byproduct of this dependency analysis is a quantitative measurement of the resilience and robustness of the system. This technique is intriguing for hardening both enterprise networks and cyber physical systems.

As a proof of concept experiment to demonstrate ADM, we have chosen the Atlassian Bamboo continuous integration and delivery tool used at Lincoln Laboratory, together with associated services, as an ecosystem whose dependencies need to be mapped. Using the time necessary to complete a standard build as one observable metric, we present data on discovering the actual dependencies of the Bamboo build server by selectively blocking access to other services in its ecosystem. Current limitations and suggestions for further work are discussed.

Keywords: Cyber dependencies · Dependency mapping
Mission mapping · Mission assurance · Software dependency analysis
Active dependency mapping

T. Bouabana-Tebibel et al. (Eds.): IEEE IRI 2017, AISC 838, pp. 169–188, 2019.
https://doi.org/10.1007/978-3-319-98056-0_7

1 Introduction

Humans are continuously expanding the complexity and interoperability of computing systems they rely on for work and daily life. In addition to desktops and laptops, we now routinely utilize multiple computing devices, many of them autonomous, e.g. smartphones, pacemakers, game consoles, programmable thermostats and cars. Daily operations also depend on data provided by external sources and the ability to reach services outside our direct control, such as GPS, Google maps, Internet Time Service, etc.

As systems become more functionally intertwined, the resulting web of complex dependencies among services becomes both increasingly important and increasingly difficult to map. The impact of degradation or compromise has an ever-increasing set of consequences in downstream system operations. Moreover, it is difficult to quantify the sensitivity of a process to the availability of other resources in its ecosystem.

However, understanding system dependencies is critical for ensuring that a cyber process is going to function as intended and when intended. Specifically in industrial, military, and critical infrastructure contexts it is essential to identify the cyber assets and capabilities that support the execution of a particular mission or task, and assess the risk to the mission or task as a function of the risks to those individual components. Mission Assurance necessitates a deeper understanding of process dependencies.

Although historically systems administrators or other Subject Matter Experts (SMEs) could be relied upon to provide the requisite expertise, this approach does not scale. The increasing complexity of today's systems make it infeasible to know or manually discover dependencies at relevant timescales. An automated methodology for discovering dependencies is needed.

2 Active Dependency Mapping Concept and Related Work

The taxonomy of dependency mapping is characterized by three dichotomies: manual vs. data-driven, active vs. passive, and network-based vs. host-based. Although it is the traditional and most common method, the manual expert-driven approach does not scale, so it will not be considered further. Data driven approaches can potentially span the other two dichotomies.

Passive approaches to service dependency detection use observations of network traffic to correlate activity between two hosts or services to infer a possible dependence [4], and these approaches are the most widely studied. Passive approaches embody several advantages. First and foremost they do not degrade the network, and hence do not generate user complaints in an operational context. In addition, from a researcher's point of view they do not require a live network to evaluate, but can be characterized with static datasets. However, they do have several drawbacks. Because they utilize correlation, they will take a long time to identify services which are mainly silent and their performance will degrade (in the sense of false alarms and missed hits) as the network becomes more congested. In addition, because correlation does not identify a direction of causation, establishing the direction of a dependency is difficult. Passive examples include Sherlock [2], Constellation [3], NSDMiner [12], Orion [9], DeDALO [8], and Methods 1–3 [6]. Of these, Sherlock, NDSMiner, Orion, DeDALO, and Methods 1–3 are all network-based, and analyze flow or packet captures. Constellation is described to be preferably endpoint-based, with each host building a correlation database. However, the authors also mention a packet-capture version, which they actually implemented to get their results.

Active approaches avoid these deficiencies by actually modifying or interfering with the distributed system and observing what happens. The "turn it off and see who complains" joke mentioned in the Abstract is a paradigmatic manual active approach. Active dependency mapping can establish causation, and will generally have a lower false alarm rate. However, because it deliberately degrades the network, it has often been thought to be "unacceptable in a production environment" [5], and as a result has been less explored than the passive variant. Previous work on active dependency mapping includes ADD [5], X-Trace [10], and Rippler [13]. ADD and X-trace are endpoint-based, while Rippler could be implemented as either a centralized delay line installed on a suitable network bridge, or an endpoint-based one (Rippler creates temporal perturbations in request timing by delaying packets with a specific pattern, which the authors refer to as "flow watermarking"). In this paper we present an experiment based on the active interference approach, and we will refer to our approach as Active Dependency Mapping (ADM) hereafter. As will be discussed below, ADM could also be implemented on the network or on endpoints, with the choice being dictated for a given network by issues of simplicity and convenience.

The situation may be summarized by Fig. 1, which presents the active/passive and network/endpoint dichotomies in matrix form. Cursory inspection yields two properties: it is top-heavy, and there are no purely off-diagonal elements. These features align with the previous observation that passive methods tend to be favored because they are less invasive, that network flow or packet collection is simpler and easier to implement than a method which requires access to the endpoints, and that active dependency mapping generally requires more complicated or intrusive modifications to the network.

2.1 ADM Concept

Active Dependency Mapping (ADM) is the discovery of service or process dependencies through artificial degradation of the network environment. The thrust of this ADM experiment is to mitigate *both* of the principal drawbacks of active approaches: their invasiveness and their need for major software modifications and/or elaborate instrumentation. Taking the two drawbacks in reverse order, let us first consider the need for custom soft- or hardware. With ADM, we draw inspiration from the joke "turn it off and see who complains," simply blocking packets from a given service and observing the effect the block. Note that blocking packets, as opposed to delaying them, is easy to accomplish with

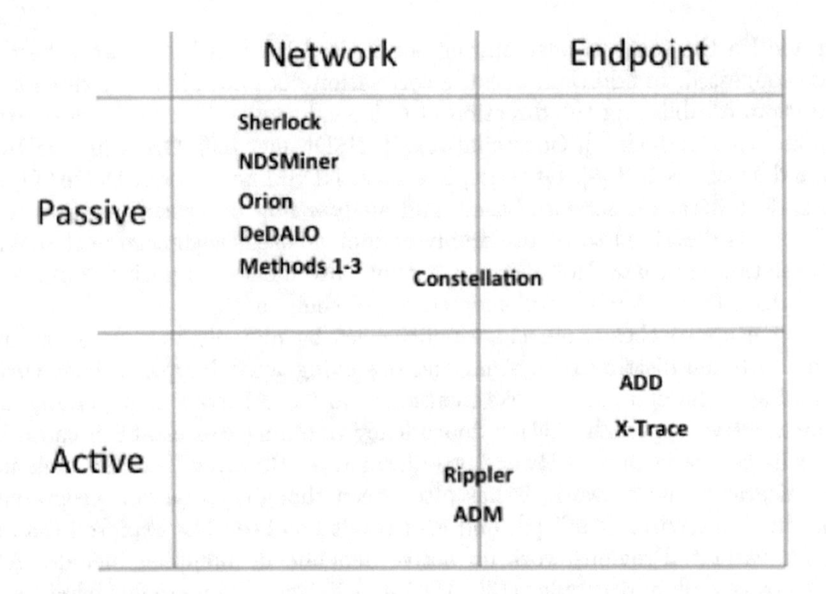

Fig. 1. Characterization of related technologies in the literature across two dichotomies: active vs. passive and network vs. endpoint.

standard techniques: changing firewall rules for a network-based approach, via the hypervisor for a VM-based system, or iptables/Windows Firewall for bare metal Linux/Windows systems. We approach the invasiveness problem by introducing the concept of Perturbative Dependency Mapping (PDM) as a refinement of ADM. In ADM, we block packets with a certain duty cycle. The goal of PDM, is in analogy with Hippocrates: "First, do no harm." Specifically we attempt to lower that duty cycle to such a degree that the service degradation is not noticeable to the majority of users, while still maintaining a high degree of statistical confidence in the results. Our experimental results will demonstrate that this goal is achievable.

Our specific contributions in this paper are the following:

- We introduce the concept of Active Dependency Mapping (ADM), which consists of simply blocking network packets from a service, and observing the results of the block.
- We note that ADM, unlike previous active approaches, is simple to implement with readily available techniques.
- We introduce the concept of Perturbative Dependency Mapping (PDM), and demonstrate with experimental data that establishing a dependency without disruption is a realistic goal.
- Finally, we demonstrate that in real-world systems dependency (and hence mission assurance) is not a static property, but is instead a function of the timescale over which it is measured.

3 Experiment Design

We wish to explicitly test the hypothesis that we can infer system dependencies by artificially degrading the network environment. Specifically we have designed an experiment to perturb traffic flow between hosts, and make observations about the impact of these perturbations on the execution of a test process. To perform live testing of Active Dependency Mapping we identify

- A live ecosystem of interdependent computers on which to experiment
- A test process requiring complex system interactions that we can discover using active mapping techniques
- A quantifiable observable used to measure the efficiency and ultimate success of the test process
- A Subject Matter Expert who can validate the dependencies we discover and alert us to dependencies that are missed

The ecosystem selected for the experiment is a collection of virtual machines that host a suite of related and interdependent software and services, specifically, an instance of the Atlassian Bamboo continuous integration, deployment, and delivery system. The functional topology of the hosts in this system is depicted in Fig. 2.

The purpose of the infrastructure is to facilitate collaborative software development, perform unit testing, and automatically build and distribute updated software versions to users. This ecosystem is sufficiently complex to exhibit unpredictable system interactions, but is also small enough to potentially test exhaustively. The test process consists of successfully modifying, building, and distributing an executable for a single code base. The explicit steps in the test process are

- Code is modified and checked into repository
- Modification detected by build server
- Update scheduled and assigned to build agent
- Executable successfully built and pushed out

Our experimental approach is to engineer a mechanism to temporarily stop the flow of traffic to and from each host, thus introducing artificial latencies in the responsiveness of various resources in the ecosystem. As discussed above in Sect. 2.1, the object of ADM is to systematically perturb traffic flow to each host in the ecosystem and observe the impact on the execution of the process. Ideally, the parameters of the traffic perturbations can be tuned so that process execution is not significantly impacted, but rather impacted in some minimal way or hardly at all, thus achieving what we call PDM.

There are three potential observables that can be used to determine the impact of the traffic outages on the test process. Most obvious is the success or failure status of the build. This observable helps quantify how disruptive the Active Dependency Mapping activities are to the successful execution of the process. However, it has two drawbacks: (1) it is a category rather than a

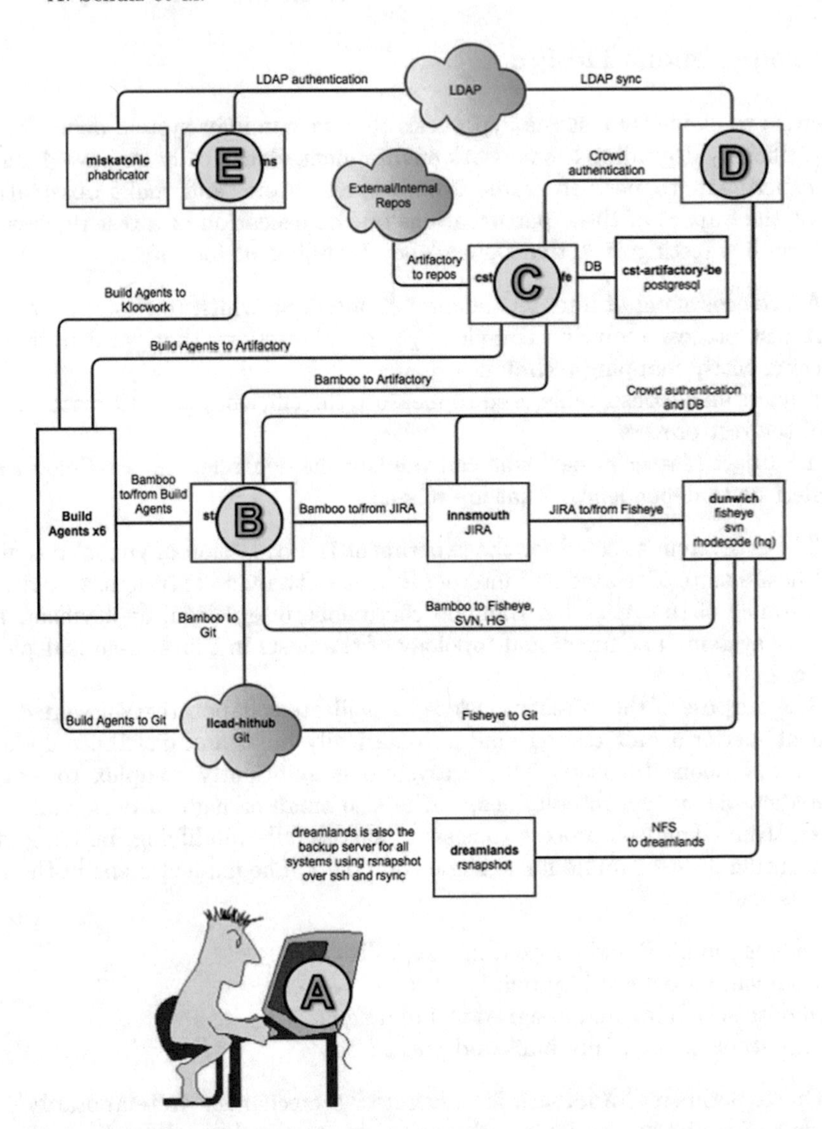

Fig. 2. An ecosystem of computers used for collaborative code development. Users at host A interface with host B, which relies on services provided by hosts C and D to successfully complete the process of modifying, building, and distributing software. Host E is not relevant to the test process in this experiment (see Fig. 3).

continuous variable, and (2) it is incompatible with our goal of perturbation rather than disruption. For successful builds, a suitable continuous observable is the time required to execute the process end-to-end. We use time as a second observable because latency is universal to many processes. Also, propagation of latencies of different durations can help quantify the sensitivity of a process to the availability of resources in the ecosystem. The third observable is log messages. When software services encounter missing dependencies, they log error messages in a log file that can be observed. Error messages are universal—most computer processes will generate log files for debugging and forensic purposes. Indeed, observing error messages to infer dependence may ultimately be preferable to observing propagation of latency, since many systems are designed to be robust to temporary degradation of network performance. For resilient systems, some dependencies may be missed if propagation of latency is the only observed indicator of dependence.

The issue of resilience is an interesting complication that arises when attempting to use active methods to determine process dependencies. Most systems, and indeed the internet itself, are designed to handle latencies and outages that naturally arise when a large number of users share a resource. For this reason, it is very important that the experiment take place on a real ecosystem of hosts, with actual users sharing the resources and variable loads in background traffic. We will enlarge on the consequences of this in Sect. 4.

Unfortunately, our desire to conduct the experiment on an actively utilized ecosystem was in direct competition with the need to provide legitimate users with a stable and reliable set of services. As a result, we performed the actual experiment on a weekend. To protect users we took time to preserve and restore the original state of the virtual machines using snapshots. We were left with about 20 h for fire-testing and data collection. To make the most of the available time, we chose to down-select the scope of the original experiment design (systematic sensitivity test of every host), in favor of gathering more statistics on a few hosts in a proof-of-concept demonstration.

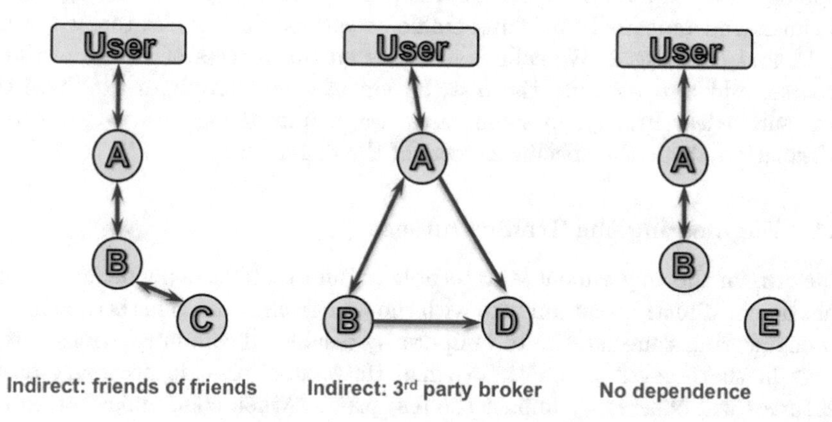

Fig. 3. Three different types of dependencies explored in this experiment. The left diagram shows a friends-of-friends dependence, the center depicts a 3rd party broker relationship, and the right shows a host with no known dependency. The letters correspond to those in Fig. 2.

The proof-of-concept experiment focuses on three different dependency types, depicted in Fig. 3. A direct dependence is one in which the host in use requires direct exchange of traffic with a host in the ecosystem (e.g. between hosts A and B in all panels of Fig. 3). A typical example of this might be a web request that is provided to the user from a web server. This simple type of dependence can easily be inferred by directly observing traffic exchanged with the host in use. However, many software dependencies do not follow this simple pattern.

A friends-of-friends dependence, shown on the left of Fig. 3, arises when a process depends on a third host (host C) that does not directly exchange packets with the user's machine (host A), but instead relays the information via an intermediary (host B). For example, a web server might retrieve data from a back end database and returns the results to the user. A second type of indirect dependence, shown in the middle of Fig. 3, we will refer to as a 3rd party broker dependency. In this more complex relationship, one host (host D) brokers a transaction between two others (hosts A and B). An example of this behavior is when a host requires the exchange of authentication information with a trusted third party (host D) before allowing the user at host A to access information or services provided by host B. Although both B and D exchange traffic with A, the successful negotiation of the process depends on host D being reachable by B. Finally, the right hand side of Fig. 3 represents a host E that is not used in the execution of our test process. Because of potential complex interactions that are possible, it's important to explicitly verify that blocking traffic to an unrelated host does not generate a statistically significant signal for any of the observables.

Our subject matter expert identified four hosts in the ecosystem that should emerge in dependency mapping for our particular test process. The letters in Fig. 2 indicate where each of the different hosts are located in the functional topology. The experiment consists of executing the test process a large number of times, and repeatedly blocking traffic at various duty cycles to each of hosts C, D, and E in Fig. 2. We collect statistics on the success or failure of the test process, and also measure the distribution of execution times for those trials that succeeded. Finally, in some cases, we examined log files to gather more information about the specific aspects of the dependence.

3.1 Engineering the Traffic Outages

The crux of this experiment is to be able to infer indirect dependencies on hosts that do not directly communicate with the user-facing host. This is accomplished by engineering a mechanism to temporarily disable all communications with the hosts in the ecosystem, one at a time. Hosts that provide necessary data or resources will observably impact the test process when communication to them is blocked.

For simplicity, we have made use of the existing firewall on the hosts to temporarily disable communications for a fixed interval. The experiment requires that the firewall repeatedly be set to reject all traffic for a fixed time interval, and then reset to default settings to resume communications as normal. To automate these changes we made use of Ansible, an open source configuration management system designed for orchestration of repetitive tasks requiring credentialed access to hosts on a network [1]. We used Ansible to update the Linux kernel firewall rules using the iptables application. During communication outages, the packet filter rules were configured to drop all packets.

The traffic outage strategy that will best identify dependencies is somewhat dependent on the test process being executed. Practitioners of ADM need to determine desired outage length (3 s in our experiment), the desired duty cycle (0%, 10%, and 50% in our experiment), and the way the outages are distributed in time (uniform distribution of start times in our experiment). Our implementation was to first generate a list of uniformly distributed outage start times, use Ansible to update iptables to block or unblock all traffic, execute a large number of trial runs of the test process to collect statistics on sensitivity to this host, and then kill the outage train at will by stopping the Ansible process script.

3.2 Experiment Day Playbook

The experiment was conducted over a weekend on an instance of the Atlassian Bamboo continuous integration system, a live ecosystem of virtual machines (VMs) that have a real population of users and a real mission to execute. It is important to note that although we had dedicated access to the ecosystem during the experiment, there were unrelated processes belonging to other users that were executing on the system during the experiment, providing an environment with a realistic level of competition for resources.

The experiment itself consisted of two asynchronous and automated processes. One automated operation is responsible for repeatedly executing the test process described in Sect. 3, and the other is responsible for orchestrating systematic traffic outages to hosts of interest as described in Sect. 3.1. The automation for the test process launched each successive trial on an interval of two minutes and thirty seconds. The automation for generating outages produced a randomly distributed sequence of outages, at either a 50% (ADM) or 10% (PDM) duty cycle (here the duty cycle defines the fraction of time that the host is unavailable).

On the day of the experiment, the ecosystem of VMs was shut down. All associated databases were backed up, and the VMs were snapshotted. Then the entire ecosystem was brought back up for the experiment. Live testing of the automated processes before experiment day was limited. Integrating the automated processes on the live systems, adjusting for differences in the host operating system, and performing fire-testing to ensure the processes operated and interacted as expected consumed approximately half of the available experiment time.

After integration and fire testing, the experiment began with a baseline measurement of sixty trials of the test process: modification of a software code followed by scheduling, building, and distributing an executable. Following the baseline measurement, systematic outages were initiated, at either 10% or 50% duty cycle. Sixty-six trials of the test process were executed for each choice of host, outage duration, and duty cycle. Three host dependency types were tested (see Sect. 3): friends-of-friends, third party broker, and a control experiment using a host with no known dependency. Unfortunately, because setup took

Table 1. A summary of host name, dependency type, and traffic outage parameter values tested during the ADM experiment.

Blocked Host	Dependency type	Outage duration	Duty cycle	Number of trials
None	None	None	None	60
Artifactory-fe	Friends-of-friends	3 s	10%	66
Artifactory-fe	Friends-of-friends	3 s	50%	66
Kingsport	Third party broker	3 s	10%	66
Kingsport	Third party broker	3 s	50%	66
Clock-of-Dreams	None	3 s	10 %	54
Clock-of-Dreams	None	3 s	50 %	0

longer than expected, data collection for the control case was cut short due to the need to restore the system to the pre-experiment state by Sunday night. Table 1 summarizes the parameters used in each of the tests.

4 Results

In this section we discuss results of an experiment to test whether artificial degradation of a network environment can be used to discover service dependencies. The hypothesis is that if a given host is utilized in the execution of a task, then blocking traffic to and from that host will result in an observable impact on task execution. In this case, the task is to schedule and execute the build of a specific code base, and we expect builds to fail, or take longer to succeed, when hosts used in the process are blocked. The central question is to determine if we can adjust the parameters of the traffic outages to achieve our PDM goal—simultaneously recover dependencies while allowing successful execution of the test process.

The instance of Atlassian Bamboo continuous integration system used in this experiment contains six independent build agents that can be used to compile and distribute code. In order to maximize the statistical samples collected over a fixed time window, we created six identical copies of the code base, and specified a particular build agent to operate on each copy. In this way, the automaton could kick off six instances of the test process in parallel. This approach also generates additional competition for resources, which is expected to introduce a scatter in the observed build times associated with the test process, rendering the results more realistic.

4.1 Timing Metric

Figure 4 shows the process execution times associated with successfully building the code base. When traffic outages were enabled they were impacting a particular host "Artifactory-fe," labeled C in Fig. 2. Artifactory is a artifact repository

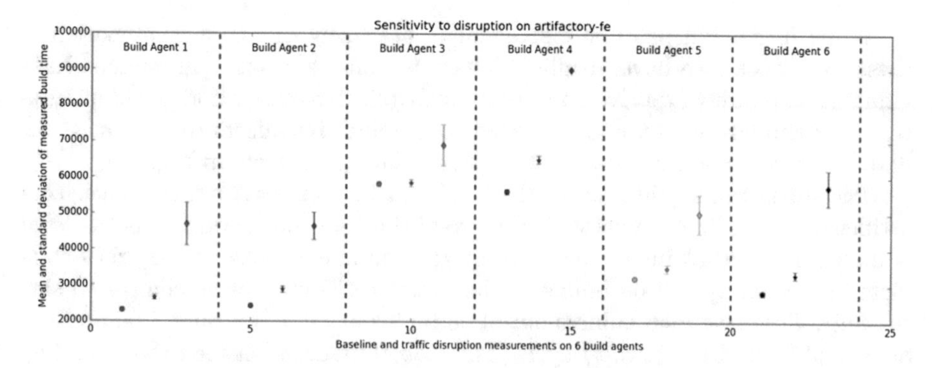

Fig. 4. Results of the experiment blocking a host with friends-of-friends type dependency. Each panel corresponds to a separate build agent. Within each panel, there are three data points corresponding to 0%, 10%, and 50% (left, center, right) duty cycle outages. The value of the y axis is the mean time to execute the test process. Error bars show the standard deviation of the mean. It is clear that more traffic disruptions lead to observable latency in the test process.

and is known to provide a friends-of-friends dependency for this test process. Each panel in the figure corresponds to a different build agent. Within each panel there are three data points: the circles (left) represent the baseline case in which no traffic outages were inflicted (i.e. 0% duty cycle), the stars (center) represent traffic outages with 10% duty cycle, and the diamonds (right) represent traffic outages with 50% duty cycle. The data points show the mean value of the build time, and the error bars show 1σ determined by the standard deviation of the mean. Any failed builds were ignored.

Figure 4 immediately suggests that the build time is observably longer when traffic to the host with a friends-of-friends dependency is blocked. Furthermore, the observed latency grows as the outages become more severe. Another interesting observation is that the baseline measurement of the build times (for which no traffic disruptions are occurring) are not consistent across build agents. We looked into this, and discovered that the Bamboo build agents were not homogeneous: build agents 3 and 4 were running Windows, while build agents 1, 2, 5, and 6 were running various flavors of Linux. This poses an issue regarding how to statistically combine the data.

The differing baselines could be due to two effects: (1) agents may require different overhead time or execute additional processes before beginning the build, and (2) the build agent machines may have different effective processing speeds. The first effect would cause an additive offset between build times, e.g. $t_1 = t_2 + \Delta_2$, where t_1 and t_2 are the build times of agents 1 and 2, and Δ_2 is a constant offset specific to agent 2. This effect could explain the large baseline build times of agents 3 and 4, the only agents running Windows. The second effect would result in a scaling factor between build times, e.g. $t_1 = \alpha_2 t_2$, where α_2 is a constant unique to agent 2.

While in general baseline discrepancies are likely due to a combination of these two effects, we have insufficient data to constrain both parameters. Individually, the purely additive and purely multiplicative corrections yield qualitative very similar results, and we opted to use additive offsets to normalize the data before statistically combining it. The results are shown in Fig. 5.

The statistical significance of the results in Fig. 5 is striking. It is clear that artificially degrading the network can reveal the dependency on this host. What is interesting is that in the low duty cycle experiment, most of the trials completed successfully. All 60 builds at the baseline 0% duty cycle completed successfully. There were 46 failures out of 66 builds at 50%, but only 7 failures out of 66 builds at the 10% duty cycle. The 9.3σ difference between the 10% duty cycle and the baseline times suggests we could block Artifactory-fe at even lower duty cycles, perhaps 5%, and still expect statistically significant results. This would allow our experiment to be less disruptive, potentially down to the level of zero build failures.

We repeated the analysis for experiments using another host "Kingsport," an authentication service known to provide a 3rd party broker dependency for the test process. This host is depicted in Fig. 2 as host D. We used the same additive corrections to normalize data across the build agents. The results are presented in Fig. 6.

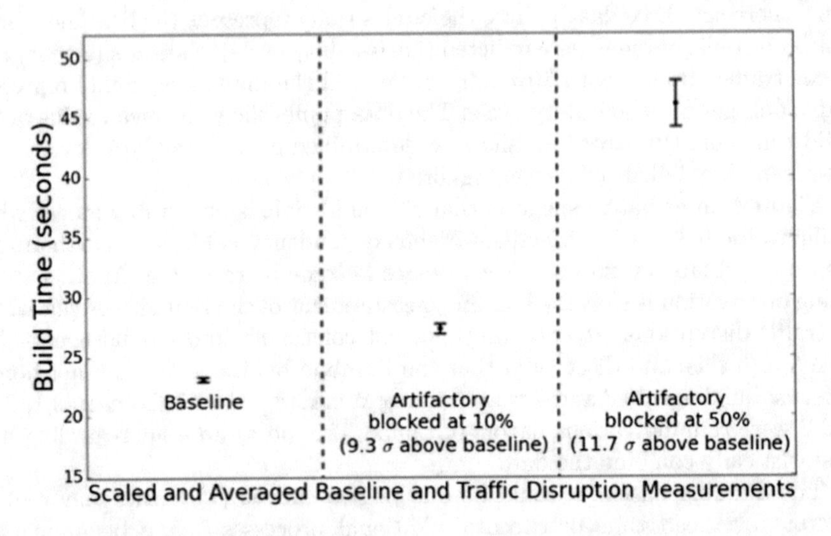

Fig. 5. The data from Fig. 4 have been statistically combined across all six build agents. There is a strong observable signature of friends-of-friends dependency in the observed latency, at both the 10% and 50%, suggesting that even lower duty cycles might be used to establish dependency.

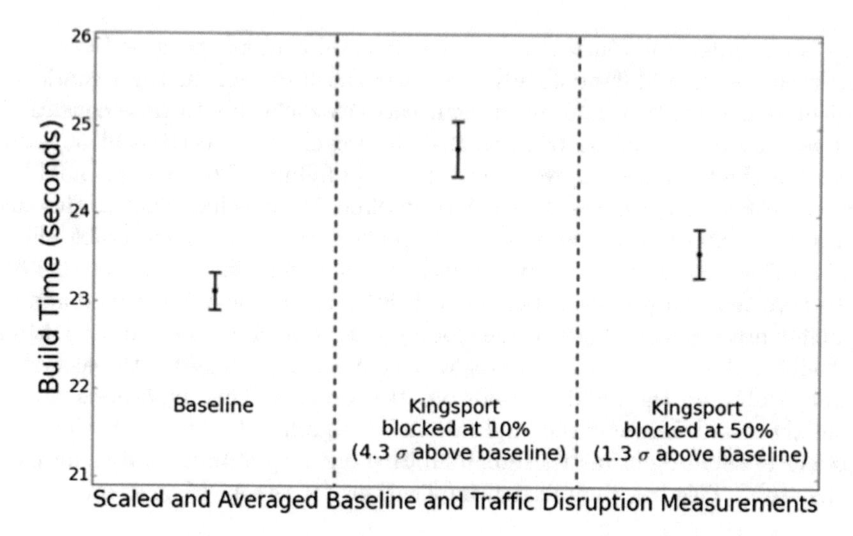

Fig. 6. A host with an expected third-party-broker dependency type does not produce an observable signature in the observed latency. Investigation revealed that the information this host was providing is cached for 15 min, providing an example of timescale-contingent dependency.

There were zero build failures across all duty cycles, and we noticed no significant, consistent increase in build time with duty cycle. This is a remarkable result. Our SME knew that Bamboo depended on Kingsport as the authentication server, yet there is no evidence that the build times were at all affected by traffic outages to it. When we looked into this in detail we discovered that the Bamboo build ecosystem is designed to cache certain information specifically to remain robust to outages. In particular, we found settings in a Bamboo configuration file to cache authenticated sessions brokered by Kingsport for 15 min. In that sense service dependency is a timescale-contingent concept. Because our experiment timescales were short, we did not see build failures or significant changes in the build times from blocking Kingsport. We would have needed to block for more than 15 min, or else employ a sparser execution of the test process (e.g. execute trials once an hour rather than once every few minutes, so that the authentication cache expires between trials), to reveal the dependency. Kingsport was not "revealed" as dependent because our testing timescales were shorter than Bamboo's caching capabilities. A different way of phrasing the result, however, would be to say not merely that the dependency was not revealed on short timescales, but that it does not exist on short timescales—dependency is not a static property, but a variable function of time.

Our final experiment involved blocking a host called "clock-of-dreams," depicted in Fig. 2 as host E. Clock-of-dreams hosts the Klocwork tool, which provides static analysis services. Since we were not requesting static analysis, according to the SME, our test process has no known dependency on this host.

This is an important control on the experimental concept, because blocking an unrelated host should (statistically) not affect the observable in a systematic way.

Unfortunately, this final experiment was cut short due to time constraints, and we only completed 54 trials at 10% duty cycle, and no trials at 50% duty cycle. For the trials we ran, we observed 0 build failures. The average build time for the 10% duty cycle was 2.7σ below baseline. We conclude that in this case, disruption of an asset not used in the test process results in no observable effects. While this is not a statistically significant result, nor did we expect this host to be revealed as a process dependency, it is unusual that builds would actually complete more quickly due to these blockages. It is possible that blocking it freed up additional resources by interfering with competing processes in the ecosystem, thus speeding up the build time for our test process. This hypothesis was not explicitly tested. However the implications are significant: when multiple human missions are sharing infrastructure, inhibiting one may improve the performance of the other. This has complicated implications for the ADM concept.

4.2 Log File Metrics

In addition to measuring the duration of a test process, we consider alternative observables for gauging the impact of traffic perturbations. Time is an intuitive and generic metric. However it is affected by competing processes and system interactions, and it relies on a test process that executes on a predictable and consistent timescale. Many tasks in the real world will not exhibit this regularity in execution time. Another generic observable that may provide a more robust indicator of disruption are the error messages in the log file associated with the test process, the motivation being that many real world software application generate log files that contain errors when expected services or resources are unavailable.

Count the Lines. In our first attempt, we do not flag on specific error messages for this test process, but instead simply count the number of lines in the log file. The hypothesis is that when outages are introduced, we will see many more error messages in the log file, and the the number of lines will increase accordingly. The number of lines in the log files from the experiment on Artifactory-fe, the host with a friends-of-friends dependence (Host C in Fig. 2) are shown in Fig. 7. The six symbol colors delineate the different build agents, and the success or failure of the build is marked with a circle (success) or star (fail).

Contrary to expectation, there are no instances of log files with significantly higher number of lines than the baseline value. Indeed, statistically significant departures all fall well below the number of lines in the baseline log files. The trials with short log files correlate perfectly with those trials that resulted in a failed build; there are error messages in those files, but the files contain fewer lines because the process truncated early, before fully executing. Error messages in successful trials were not so numerous as to provide an observable contrast against normal fluctuation in the number of lines in the log. We conclude that number of total messages do not make a good generic observable.

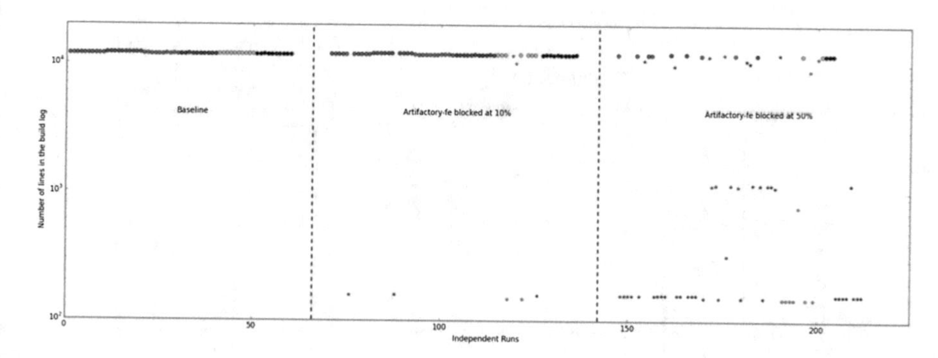

Fig. 7. Each point in this plot corresponds to one trial of the experiment on Artifactory-fe, a host with friends-of-friends dependency of the test process. The six colors delineate the separate build agents. Circle symbols note the successful trials while stars note the trials that failed. The y-axis shows the number of lines in the log file associated with the test process. Trials that fail actually produce many fewer lines because the process truncates early. However, trials that succeed but are delayed compared to the baseline case do not exhibit an observable increase in the length of the log file due to the generation of error messages. This observable is insufficiently sensitive to use as a perturbative mapping observable.

Natural Language Processing. Our second attempt focuses on the content of the log files themselves, rather than merely their content. We use techniques borrowed from Natural Language Processing (NLP) to search for indicators of failure and delay within the build logs. Our hypothesis is that certain words or expressions would be written to the build logs if any of the component machines were experiencing minor outages or delays. In that case, bag-of-words topic modeling should be able to automatically identify which build logs contain delay messages, and we would not need to rely on time measurements to make that distinction. We begin by converting each build log into a list of its component words (or unigrams), ignoring all characters except English letters. From these lists, we compute a term-frequency inverse-document-frequency (TF-IDF) matrix. Each column of this matrix represents one of the unigrams in the vocabulary of the entire corpus, and each row represents one of the build logs. For each term-log pair, we compute the term frequency (1 if the term appears in the log, 0 if not) and divide it by the logarithm of the fraction of all build logs that contain the term. The inverse-document-frequency component de-emphasizes overly common words, and we use binary term-frequency instead of word counts so that we don't lend too much weight to the longest logs in the corpus. As the TF-IDF matrix is very large and sparse, we reduce its rank using a process called Latent Semantic Analysis (LSA), which shares strong similarities with Principal Component Analysis. Each row of the LSA matrix still represents a single build log, and each column represents a linear combination of unigrams, building a semantic space that efficiently characterizes the variance in the original TF-IDF matrix. Figure 8 illustrates some important results from the LSA. In this plot, each data point represents a single build log, the text of which has been

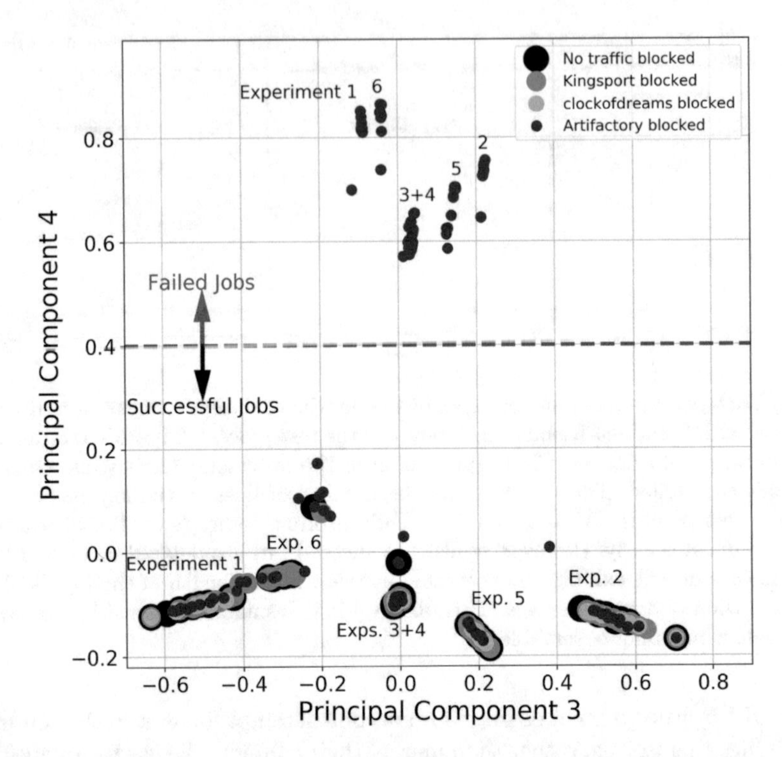

Fig. 8. A two-dimensional slice of the semantic space created by performing LSA on the build logs. Each data point represents a build log, colored according to the traffic that was blocked during that particular job. Marker sizes vary only to improve the visibility of overlapping points. The fourth component of the LSA effectively separates failed and successful logs, and data points tend to cluster with points from the same experiment, but we found no means of automatically identifying the logs from delayed builds with this technique.

projected onto a two-dimensional slice of semantic space. These dimensions are the third and fourth columns of the LSA matrix ("Principal Components" 3 and 4), and they are plotted along the x- and y-axis, respectively. Principal component 4 clearly separates the failed and successful jobs. Within both of these separated group, data points are further clustered according to the experiments that produced the logs. This is because each experiment performed builds on hosts with different operating systems and configurations, so the build logs from different experiments contain different words and expressions. Ideally, for perturbative mapping we hope to identify an NLP differentiator between cases that were and were not systematically delayed due to traffic outages. Unfortunately, in this set of experiments we find no principal component that could be used to identify the logs of builds that were merely delayed, rather than failed. Thus, this NLP experiment does not by itself get us to a Perturbative Dependency Mapping. However, by appropriately reducing delay time, it should be possible to reduce the number of build failures to a very low and perhaps tolerable fraction (say less than 5%).

5 Discussion and Lessons Learned

5.1 Application to Cyber Mission Assurance

Up to this point we have discussed ADM in the context of dependency mapping, but the general concept can be extended to apply to cyber mission assurance. In both military and industrial contexts, it is important to identify the cyber assets and capabilities that support the execution of a particular mission or task, and assess the risk to the mission or task as a function of the risks to individual components. The military has spent more time and effort than the civilian sector studying the mission assurance concept, so we will quote from their definition:

> A process to protect or ensure the continued function and resilience of capabilities and assets – including personnel, equipment, facilities, networks, information and information systems, infrastructure, and supply chains – critical to the performance of DoD [Mission-Essential Functions (MEFs)] in any operating environment or condition [7]

Cyber mission assurance considers the cyber aspects of mission assurance in general, and focuses on both identifying and making resilient those cyber assets critical to Mission Essential Functions. These critical cyber assets are often termed" cyber key terrain (CKT)" by analogy to key terrain in physical military operations. ADM is clearly a valuable method of both identifying and evaluating the resilience of CKT; we shall elaborate on this point later in the discussion.

5.2 Potential for Operational Deployment

There are many practical lessons learned in this research that inform how an ADM approach might successfully be deployed in an operational context. One of the most important insights arising from this experiment is that the concept of cyber key terrain, i.e. the computer systems and networks that directly support the execution of a mission or task, is inherently time dependent. We saw an example of this in the experiment, when a host required for the exchange of authentication information did not appear in the observed key terrain because credentials had been cached. On time intervals longer than 15 min, this host is most certainly in the key terrain, but the process is robust to outages of short duration that occur after credentials have already been exchanged. The experiment we designed was not sensitive to the dependency, but the same experiment executed over longer timescales would certainly have been. Active Dependency Mapping techniques have inherently built in timescales, and only dependencies relevant on the probed timescales will be identified. For operation deployment, then, we conclude that the final implementation of an ADM dependency sensor should investigate dependencies on a variety of timescales by adjusting outage duty cycle and duration, as well as the time interval associated with executing the test process. It is worth noting that the ADM approach offers unique insights on timescale not available with other dependency mapping technologies.

Another important lesson for operational deployment is that directly manipulating the firewall on each host is logistically very challenging in a heterogeneous environment with a diversity of operating systems. In this experiment we spent a lot of time adjusting the code responsible for generating outages on each host to accommodate the heterogeneity of the ecosystem. For operational deployment, this level of boutique tinkering for each host would not be scalable. To use this technique operationally would require either that the hosts in the ecosystem be more standardized or that the traffic outages be generated at a higher level: in the hardware, potentially at router or switch locations, or at the hypervisor interface for a virtualized ecosystem.

Finally, there is a strong need to develop a broader range of effective and generic observables and metrics to serve as indicators of dependency. Latency propagation was effective at identifying dependency in some but not all cases, and it suffers from a couple of drawbacks. First, the experiment as we designed it measures resiliency to disruption more than it measures actual dependency. This is not surprising: the strategy is very similar to the well known member of the Netflix Simian Army, the Chaos-monkey [11], designed to ensure the resiliency of the product. But propagation of latency is flawed as an observable for perturbative dependency mapping because it sometimes induces failure, while in other circumstances the process is robust to latency. In other words the impact of latency is likely to be very mission and process dependent, which makes it less generic an observable than we first had hoped. This is not to say that latency should not be used in an operational context. First, for some dependency types it will work well, and second, it serves to evaluate resilience, which is an imperative in the Mission Assurance context. Rather, an operational ADM architecture would be well designed to employ an array of generic observables to infer dependency, rather than a single observable. The primary challenge lies in identifying other quantities that can be easily measured to gauge mission impact.

6 Conclusions

In this paper we have argued that automated methods are needed for effectively determining the dependencies of processes on other assets and resources in the ecosystem. We present a brief taxonomy of existing methods, and introduce Active Dependency Mapping, a technique that involves systematically blocking traffic to and from hosts in a network environment in order to directly detect observable impact on the execution of a test process. We suggest that it may be possible to artificially degrade the network *perturbatively*, so as to generate observable phenomena that identify a dependency without noticeably disrupting the user or the task. This approach can be applied to enterprise networks or cyber-physical systems.

We present the results of an experiment conducted on a real world ecosystem of assets used at the laboratory for collaborative code development and software deployment. We have defined a specific test process, and observed the propagation of latency generated by introducing systematic outages of traffic to other hosts in the ecosystem. The proof of concept experiment shows definitively that some types of dependencies can be captured with artificial network degradation. Given the statistical significance of the results, it is likely that this will remain true even at very low levels of degradation, which would allow for zero failures of the test process. Other dependencies were not identified, but we learned how the experiment could be modified to detect them in the future.

Preparation for operational use requires additional research. First we should leverage traffic outage parameters to explore dependencies at different timescales. Second we should identify a wider variety and more generic mission impact observables, rather than depending on the single observable tested in this proof-of-concept experiment. Moving away from direct firewall manipulation will help reduce the complexity of implementation for more seamless deployment. Finally, it is important to validate efficacy for a variety of expected use cases. These preparations will be the subject of future research.

In conclusion, Active Dependency Mapping techniques exhibit strong advantages. In particular, they can be entirely data driven. They have the potential to infer existence of complex and indirect system dependencies. And finally, they can probe dependency at different relevant timescales, thereby contributing to the evaluation of cyber Mission Assurance.

Acknowledgements. We are very grateful to Robert Cunningham and Dinara Doyle for access to the networked ecosystem where we performed our experiments. We would also like to thank Jeffrey Gottschalk and Martine Kalke for their support.

This material is based upon work supported under Air Force Contract No. FA8721-05-C-0002 and/or FA8702-15-D-0001. Any opinions, findings, conclusions or recommendations expressed in this material are those of the author(s) and do not necessarily reflect the views of the U.S. Air Force.

References

1. Ansible it automation (2016). https://www.ansible.com/it-automation
2. Bahl, P., Chandra, R., Greenberg, A., Kandula, S., Maltz, D.A., Zhang, M.: Towards highly reliable enterprise network services via inference of multi-level dependencies. In: ACM SIGCOMM Computer Communication Review, vol. 37, pp. 13–24. ACM (2007)
3. Barham, P., Black, R., Goldszmidt, M., Isaacs, R., MacCormick, J., Mortier, R., Simma, A.: Constellation: automated discovery of service and host dependencies in networked systems. Technical report, MSR-TR-2008-67 (2008)
4. Bartlett, G., Heidemann, J., Papadopoulos, C.: Understanding passive and active service discovery. In: Proceedings of the 7th ACM SIGCOMM Conference on Internet Measurement, pp. 57–70. ACM (2007)

5. Brown, A., Kar, G., Keller, A.: An active approach to characterizing dynamic dependencies for problem determination in a distributed environment. In: 2001 IEEE/IFIP International Symposium on Integrated Network Management Proceedings, pp. 377–390. IEEE (2001)
6. Carroll, T.E., Chikkagoudar, S., Arthur-Durett, K.: Impact of network activity levels on the performance of passive network service dependency discovery. In: Military Communications Conference, MILCOM 2015-2015 IEEE, pp. 1341–1347. IEEE (2015)
7. Carter, A.: Mission assurance strategy. Technical report, Department of Defense (2012)
8. Casalicchio, E.: Dependencies discovery and analysis in distributed systems. In: International Workshop on Critical Information Infrastructures Security, pp. 205–208. Springer (2011)
9. Chen, X., Zhang, M., Mao, Z.M., Bahl, P.: Automating network application dependency discovery: experiences, limitations, and new solutions. OSDI **8**, 117–130 (2008)
10. Fonseca, R., Porter, G., Katz, R.H., Shenker, S., Stoica, I.: X-trace: a pervasive network tracing framework. In: Proceedings of the 4th USENIX Conference on Networked Systems Design Implementation, p. 20. USENIX Association (2007)
11. Izrailevsky, Y., Tseitlin, A.: The netflix simian army. The Netflix Tech Blog, July 2011
12. Natarajan, A., Ning, P., Liu, Y., Jajodia, S., Hutchinson, S.E.: NSDMiner: Automated Discovery of Network Service Dependencies. IEEE (2012)
13. Zand, A., Vigna, G., Kemmerer, R., Kruegel, C.: Rippler: delay injection for service dependency detection. In: IEEE INFOCOM 2014-IEEE Conference on Computer Communications, pp. 2157–2165. IEEE (2014)

Author Index

T. Bouabana-Tebibel et al. (Eds.): IEEE IRI 2017, AISC 838, p. 189, 2019.
https://doi.org/10.1007/978-3-319-98056-0

Printed in the United States
By Bookmasters